BRÜSSEL
London
Stärke
Reichs
DEN INVASIONSTERMIN
Churchill
CW01646472

GERMAN HELMETS OF THE SECOND WORLD WAR

By Branislav Radovic

HELMETS OF THE FIRST WORLD WAR *(with Michael Haselgrove)*

GERMAN HELMETS OF THE SECOND WORLD WAR: Vol. 1

GERMAN HELMETS OF THE SECOND WORLD WAR: Vol. 2

GERMAN HELMETS OF THE SECOND WORLD WAR

1

Branislav Radovic

To Greg
with thanks for
all your help
– I hope it provides
some fun!

March 2002

Schiffer Military History
Atglen, PA

Acknowledgments
The author wishes to thank the following for their particpation: Marcus Cotton, Malcolm Fisher, David Hucklebridge, Larry Klinger, Brian Maederer, Greg Nevill, Bill Shea, Barry Smith and Ellwood von Seibold.

Dedication
This volume is dedicated to my darling wife Anna Chajkin.

Book Design by Branislav Radovic and Robert Biondi.
Photography by Robert Biondi.

Copyright © 2002 by Branislav Radovic.
Library of Congress Catalog Number: 2001094917.

All rights reserved. No part of this work may be reproduced or used in any forms or by any means – graphic, electronic or mechanical, including photocopying or information storage and retrieval systems – without written permission from the copyright holder.

"Schiffer," "Schiffer Publishing Ltd. & Design," and the "Design of pen and ink well" are registered trademarks of Schiffer Publishing, Ltd.

Printed in China.
ISBN: 0-7643-1447-5

We are always looking for people to write books on new and related subjects. If you have an idea for a book, please contact us at the address below.

Published by Schiffer Publishing Ltd.
4880 Lower Valley Road
Atglen, PA 19310 USA
Phone: (610) 593-1777
FAX: (610) 593-2002
E-mail: Schifferbk@aol.com.
Visit our web site at: www.schifferbooks.com
Please write for a free catalog.
This book may be purchased from the publisher.
Please include $3.95 postage.
Try your bookstore first.

In Europe, Schiffer books are distributed by:
Bushwood Books
6 Marksbury Ave.
Kew Gardens
Surrey TW9 4JF
England
Phone: 44 (0)208 392-8585
FAX: 44 (0)208 392-9876
E-mail: Bushwd@aol.com.
Free postage in the UK. Europe: air mail at cost.
Try your bookstore first.

CONTENTS

A Brief History 6

M1916-18 21

M1932 79

M1935 89

M1940 167

M1942 247

M1942/45 323

A BRIEF HISTORY

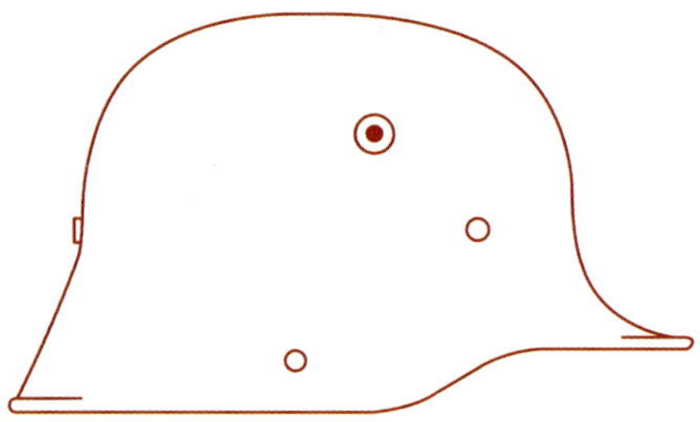

THE MODEL 1916

The M16 steel German helmet was designed to replace the leather, steel and felt versions of the pickelhaube because of the increasing number of head wounds sustained by the German army.

Experimental research for the helmet, under the direction of Professor Schwerd of the Technical Institute in Hanover, began in mid- to late 1915 with the creation of various shapes and configurations. While the order in which these versions were produced is unclear, the earliest design sketches and patent drawings establish that the horn-like ventilation lugs, engineered to carry the brow plate, were always an integral part of the design.

The production version of the M16, first issued to troops fighting at Verdun during the winter of 1915-16, was available in six sizes: 60, 62, 64, 66, 68 and 70. Sizes 60 to 64, because of their smaller dimensions, incorporated a pair of two-piece extended ventilation lugs of variable thickness to hold the one-size brow plate. Several companies manufactured the M16, among them Eisenhuttenwerke and Siemens-Halske. The factory's code was usually found on the inside left earguard followed by the helmet shell size. For example, the inscription for a size-64 produced by Eisenhuttenwerke, the most prolific manufacturer, is ET64 and that for Siemens-Halske, Si64. Other factory prefixes include K, W, TJ, BF and G.

The M16 is distinguished by its thick leather liner band onto which are stitched three leather two-tongued tabs (three-tongued in the patent drawings), They each hold a pad of horsehair or cloth to cushion the head, one at the forehead and one near each ear. The wearer corrects the position of the pads by drawing together the tongues with a strip of leather or lace.

The tabs are usually brown, although a light, off-white colour often appears in the M17 and M18. Chinstraps on both the M16 and M17 vary in width and thickness but, in regulation configuration, were always attached to the helmet shell by clips that locked over the Model 91 lugs (as used on pickelhaubes) on the inside of the earguard. However, chinstraps frequently broke through wear and tear and were often repaired or replaced in the field. The strap, which is adjusted by using the two sliding buckles, is joined to brass or steel lug clips at each end and stitched or single-riveted into place. Problems were encountered in keeping the straps attached to the lugs as evidenced by examples showing how the wearers have beaten over the edges of the lugs in an effort to secured the straps permanently.

The brow plate is heavy, weighing about 5 lbs (2.5kg). It is bullet-proof up to 150 feet (50 metres) and secured to the helmet by a leather strap with a double rivet on each side, a configuration also used in the later M18. The strap is tightened and fastened by a prong buckle in a series of strap holes that often extend, for reasons now unknown, much further than could reasonably be used. The double-rivet attachment system, however, did not allow enough movement of the leather, with the resulting stress causing the leather to fail. Indeed, this is perhaps one of the reasons why original straps on M18s have only rarely survived.

The quality of the finish to the brow plates varies greatly; often the edge is not smooth where it curves over the helmet, and, indeed, marks from the rollers and files used to produce the plate are often still plainly visible. They were usually painted in the same grey-green as the helmets, but there are many examples where a much darker paint was used but, because the paint layer was so thin, now appear to have been unpainted. The brow plate was issued when required and then returned by the soldier when not needed. As brow plates passed from helmet to helmet, it is most unlikely that any would have been painted to match an individual helmet's camouflage.

Manufacturers' markings are usually stamped into the metal on the inside right. Krupp, for example, is represented by three overlapping circles. Another frequent mark is the letter R followed by either numerals, a bell shape or a series of letters and numbers. Also frequently found are the letters AK stamped in ink, which is the official military acceptance mark. This ink stamping is also found on body armour that was sometimes worn at the same time as the brow plate. However, some fully genuine brow plates have no markings at all.

Austrian brow plates – which are even rarer than their rare German counterparts – are usually almost identical in shape and colour, but some are of a thicker gauge metal and have a finer finish to edges. The Austrian strap is made of webbing and attached by a single rivet at each side of the plate or by stitching. The same type of strap is used on Austrian helmets, having metal reinforced holes for the prong buckle and leather or leathercloth reinforcing at the strap-end.

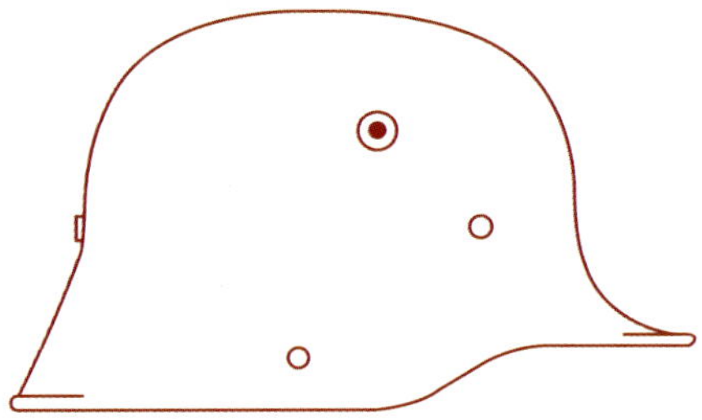

THE MODEL 1917

The M17 is a variation of the M16. In the spring of 1917, by which time leather was in short supply, a 2.6cm-wide metal liner band was introduced. It is attached to the helmet shell in the same way as the M16 leather band and has a lower edge rolled over and is joined at the back of the shell. It is perforated by three groups of 32 small holes where the two-tongued leather liner sweatpads are placed. These holes seem to be for sewing the pads to the liner band, even though the rolled edge of the liner holds the sweatpads in position, making sewing unnecessary.

The brown leather sweatpads and the horsehair cushions are similar to the M16's but a lighter, off-white leather sweatpad appears increasingly. This lighter leather, which can take on a greenish tinge in certain light conditions, is more supple than the M16's brown type. It is nearly always found in better condition than the brown and seems not to promote the same rusting reaction with the metal liner band that the brown does.

The M17 sweatpads, like the M16's, have a piece of hessian or cotton sewn along the bottom edge and each side, usually with tucks at the corners, to form a pocket for a padded cushion constructed from two pieces of cotton, usually plain but sometimes patterned. The padding is held in place by four deep-buttoned stitches across the middle. Care was taken to gather and seam the edges on both sides. The cushions are held in place by a strip of loose- or tight-weave cotton tied in a knot across the back of the sweatpad.

The M17 chinstrap is also the same as the M16's and attached to the shell via M91 lugs in the same way as its predecessor. The shell maker's mark and size stamp remain on the inside left earguard. Apart from the introduction of the metal liner band and the sweatpad innovations, the M17 is little changed from the M16.

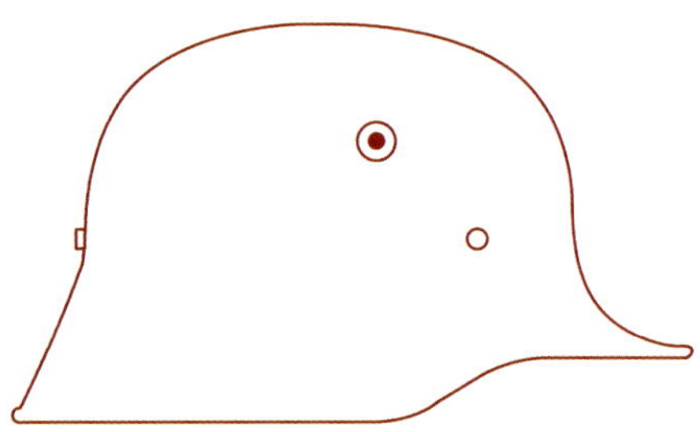

THE MODEL 1918

With the development of the M18, Germany entered the age of the modern helmet. The M91 chinstrap lugs – a hangover from the 19th-century pickelhaube era – were abandoned, probably because the M16 and M17's straps frequently fell off. They were replaced by an arrangement that directly attached the straps to the liner with rectangular swivel rings on each side in front of the ear.

The new chinstraps were made of leather and in two main parts. The longer right-side strap is looped through its swivel ring and has a sliding buckle. A grey carbine clip is sewn at the end. The shorter left-side strap is fastened to its swivel ring by a short loop and secured with one or two grey or black rivets similar to those attaching the brow plate's straps. At the end of the short strap is a sewn-in, grey rectangular ring to receive the carbine clip and a leather flap to protect the wearer's skin from chafing on the metal parts when the strap was joined. The flap's smooth-skinned side usually has impressions of the maker's mark and the date of manufacture. Examples include "Hermann Freimann 1918 Nürnberg" and "Ludwig Raffael Nurnberg 1918."

As mentioned above, it is unusual today to find an M16 or M17 retaining its original chinstrap. Unfortunately, the problem is even greater with regard to the M18, a state of affairs likely due, at least in part, to an important design flaw and a shortage of war materials. The rivets that attach the shorter strap are so close to the liner rings, the strap is prevented from swivelling easily, thereby forcing the leather on the other side of the rivets to carry most of the flexibility. Moreover, the leather available at this late stage in the war was far less supple than that used before. Germany's under-funded, post-war army found carbine clips useful in many ways. For example, they were used on the chinstrap of the post-war transitional army helmet with the M27 leather liner and on the M36 fallschirmjäger (paratrooper) helmet, which used four clips.

The M18 liner's sweatpads were apparently more often made of the off-white leather than of the brown, but, where the brown sets still exist, each pad is impressed with the maker's name, which are sometimes the same as found on the chinstrap.

The paint on the M18 is usually dark green, sometimes almost black, though the latter is most likely the result of the 80-year ageing process. However, there are examples with the Army's standard matte grey-green sprayed finish, and others with a brushed surface that sometimes is shiny, and on occasion, displays runs due to overpainting.

(collectors' note) One must not forget the M18 is more than 80 years old and therefore should show signs of age, such as a dark brown/black smooth dormant rust on the edges where the paint has chipped as well as dark rust patches in chafed areas with clearly defined areas of paint and rust.

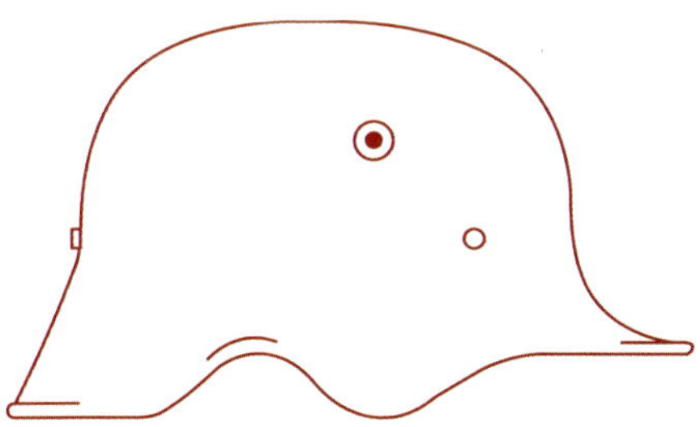

THE MODEL 1918 CUT-OUT

The M18 cut-out is, apart from a single important exception, no different than the standard M18.

In August 1918, when the German War Ministry decided that the M91 chinstrap needed replacing, it had also received reports that troops in the trenches had complained that the M16/

M17's low earguards impaired their hearing. Consequently, it ordered the production for evaluation of 2,100 size-64 helmets with a section cut out near each ear. The innovation was not only generally successful but also allowed a field telephone operator to work without removing the helmet. The cut-out version then became favoured during the inter-war years by cavalry as well as infantry officers.

Many cut-outs were covered with a fibrous texture called "flocking" designed to produce a permanent matte finish, because the War Ministry also discovered in 1918, that the helmets became shiny through use and dangerously reflective, especially in moonlight. The majority of surviving cut-outs are unflocked, although this might be the result of the flocking being removed to apply camouflage.

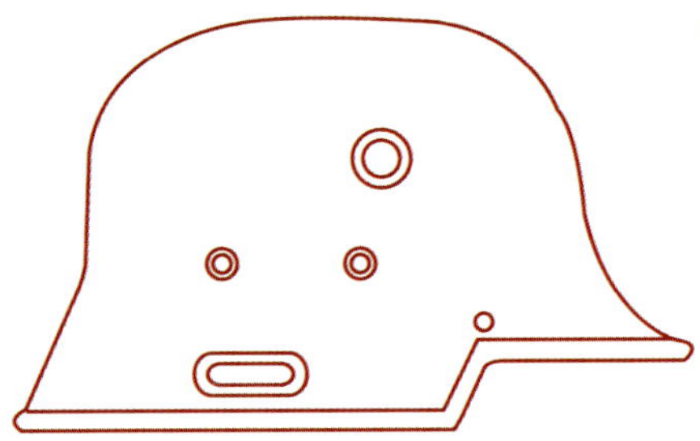

THE INTER-WAR PERIOD 1919-38

By the end of the war, Germany had produced 7.5 million helmets. Many thousands were, of course, taken home in Allied knapsacks to France, Britain and America, continuing the ancient tradition of displaying battlefield prowess by the possession of the enemy's armour.

The Treaty of Versailles, imposed by the Allies, required Germany to reduce its army to a maximum of 150,000 men, to having its airforce dismantled, its warships impounded and its colonies carved-up – and to provide raw materials in reparation. In short, the immediate post-war period in Germany was marked in military terms by a shrinking and consolidation of its armed forces and their equipment.

Thousands of the remaining, unused helmets were smelted down, partly for German use but also some of it going to the equally steel-needy victors. However, some surviving surplus helmets were sold off by the German War Ministry at DM19.75 each.

During the 1919 Spartacist revolution, ex-servicemen reformed into combat groups reflecting polarising political views, generally on the lines of a communist-influenced left and a Freikorps right. Contemporary photographs show M16, M17 and M18 helmets being worn by both sides, but mainly by Freikorps units. They show the special visorless M18 – produced for Germany's Turkish allies but never sent – being worn by Freikorps men fighting on the streets of Berlin. Wartime surplus once more was being recycled over and over again.

In the meantime, Germany's remaining regular forces were required by the government to claim all the helmets they would need by 1 April 1920, with the M18 cut-out being earmarked for the cavalry. All unclaimed helmets were to be sold off. All German army (Reichswehr) helmets traditionally had no official insignia but in

1920 provincial shields began to be painted on either one or both sides under the ventilation lugs. By 1920, Freikorps units were adopting their own insignia, many being hand-painted onto the front of their helmets. In January 1924, the Reichsmarine adopted a helmet transfer (decal) of a yellow cross-anchor device on a white shield bordered in yellow to be applied on the left side under the ventilation lug.

The war-time colour scheme of grey-green paint was also changed to a slightly more yellow shade, now known as apple-green, that was usually applied by brush rather than by spray. Helmets for the Reichswehr were in short supply and issued "as needed" and returned to stores after use. A regulation, dated 12 November 1927, required officers to pay a "wear and tear" charge on the return of their helmets.

Throughout the inter-war period, some wartime manufacturers continued to produce M18-style helmets in relatively small numbers. They are often made in a heavier gauge steel and sometimes show a slight crease or ripple across the visor. They are fitted with M18-style liners made from various materials, construction and quality. A new liner was produced in 1927, still with three pads but attached to each other and more securely attached to the metal liner band.

Another liner system was produced in 1931 (M31) to fit easily into M16, M17 and M18 shells via the usual split cotter pins. While similar to the M18 liner system in that the chinstrap attached directly to the metal liner band, it has a six-, eight- or nine-tongued leather lining, pierced for scalp ventilation. It is fully adjustable using a drawstring and was manufactured in 11 sizes – 53 to 61 inclusive. The chinstrap is a simplified version of the M18's carbine clip and later a simple prong buckle on one strap and a plain leather strip with 13 holes for adjustment on the other. The buckle was usually placed on the left to avoid interfering with the sighting and firing of rifles by right-handed men. It could be switched to the right side for left-handers.

In 1932, a re-design of the combat helmet was ordered for an increasingly confident German army. The war-time helmets were considered too heavy, too restrictive audibly and visually, and a primary cause of heat-stroke. A total of 510 of the helmets (M32) were produced for testing by infantry, communications and artillery units. The lighter design sacrificed protection for greater comfort. It was made from the pressed compound Vulkanfiber, which, while extremely light, required the rim to be reinforced with a pressed metal band to retain rigidity. The liner and chinstrap assembly was far superior to the M18's and M27's. The shell was a cut down version of the M18's, providing better vision for taking aim, and less restriction for the wearer and therefore far more comfortable. A slot was positioned on each side of the rim, apparently for its attachment to pack or belt. The original insignia was a metal shield carrying the national colours and attached to the left side by two holes in the shell, but this was abandoned during testing in favour of what would soon become the standard transfer – the tri-colour national shield (diagonal sections of black, white and red) to the right and the eagle and swastika to the left. Tests continued well into 1934 but were cancelled due to the imminent arrival of the Model 1935.

In 1933, the Reichsmarine's crossed anchor insignia was replaced by the tri-colour (red/white/black) national shield. When transfers were not available, the insignia was hand-painted. The army also adopted this configuration, which replaced all provincial insignia. Upon Hitler's inauguration as chancellor in 1933, the national shield was moved to the right side of all helmets and a black shield bearing an eagle atop a swastika in silver for the army, and in gold for the navy, was placed on the left side. The airforce, still officially non-existent, did not adopt its own insignia until 1935.

The Schützstaffel (SS) went through its own convoluted evolution of helmet insignia. The first official version was a black circle outlined in white, with twin SS runes in black also outlined in white, and positioned on the right side under the ventilation lugs. On the left was a white-outlined black swastika. In February 1934, Hitler's bodyguard, the Leibstandarte SS Adolf Hitler (LSSAH), adopted a new set of insignia. On the right were silver SS runes set within a white-bordered black shield with its top-left corner clipped. On the left was the national shield used by the army and navy. However, within a year the LSSAH replaced its right-hand runic shield with a black-bordered unclipped silver shield with black SS runes inset. The SS Verfugungstruppe, later renamed the Waffen (armed) SS, replaced its SS runes within a circle and swastika with a black-bordered unclipped silver shield with black SS runes on the right side and a black-bordered shield on the left bearing the new state colours of a red background with white circle containing a black swastika.

By 1935, this configuration had also been adopted by the LSSAH when it became part of the Waffen-SS. Also by this time, SS helmets in general – which before 1934 were usually painted black – were being issued in apple- or grey-green in line with the armed SS's military profile. Black versions were reserved for guard/parade use.

The RZM SS helmet is another of note from the early 1930s. It is an M18-style helmet specifically made by the Reichszeugmeisterei (RZM), the Third Reich's equipment office, to increase the number of helmets for the SS, which had successfully argued that it was not receiving its fair share.

The helmet is sometimes lighter than the M18 and stamped on the inside of the shell near the rim below the ventilation lugs. The RZM logo is on the left; SS runes within a circle are on the right. The M18-style liner is permanently attached to the shell by single-core rivets instead of the usual split cotter pins. This helmet, which was painted either very dark blue-grey or black, carried no insignia.

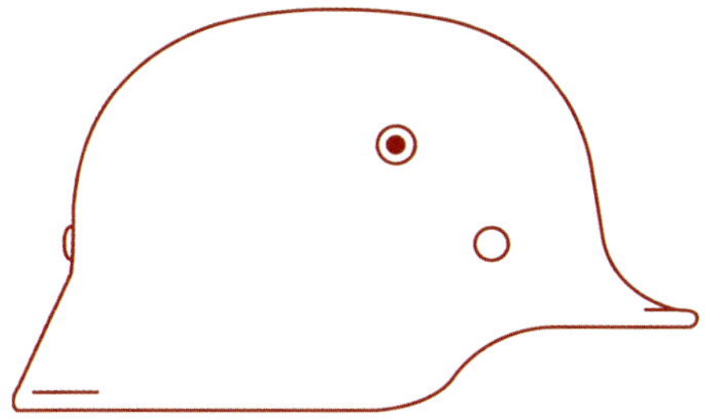

THE MODEL 1935

As mentioned earlier, a Vulkanfiber prototype aimed at replacing the wartime M18-style was tested in 1932 and then rejected, but, in 1935, a completely new model of steel helmet was designed and approved – the Model 1935 (M35). At first appearance it is simply a cut-down version of its immediate predecessors, without the prominent ventilation lugs for the now-redundant brow plate. The M35's innovations were of great importance in the development of the 20th-century helmet.

The ventilation holes on the newcomer were less obtrusive, having a hole punched through the helmet shell on both sides with a bushing (metal grommet) inserted to create a smooth finish. The edge of the helmet shell was rolled inwards, again to create a smooth finish, as in the M16, 17 and 18.

With superior steel and steel drawing methods now available, came the possibility of a more compact, rugged, safer and more comfortable helmet, befitting the birth of the new, modern German army (Wehrmacht). A painstaking period of issue and replacement took several years to complete. There were never enough of the new style M35s to go around.

The M35 shell was produced in six sizes – 60, 62, 64, 66, 68 and 70 – and was fitted with the M31 liner, which was adjustable through 11 sizes. The split cotter pins which held the liner in the shell had to be redesigned because they proved to be too weak and prone to shearing. Many early M35s found their way to China to equip Chang Kai Shek's nationalist army.

Applying insignia by transfer continued. Army and navy helmets were finished in a smooth apple-green, with the army's silver eagle and the navy's gold one on the left side under the vent. Both have the national shield on the right. The original airforce M35 also has an eagle on the right and the national shield on the left, but is finished in blue-grey and the eagle depicted in-flight , the so-called "droop tail" pattern. Within a year, however, the airforce eagle was redesigned. It remained in-flight but was slightly larger and more aggressive in appearance.

The SS version carries the familiar black-bordered silver shield with black SS runes on the right and black-bordered red shield with black swastika on a white circle on the left. The helmet was either in the apple-green for the field or the ceremonial black, the latter often the result of overpainting the former.

All M35s, similarly to their WWI predecessors, were stamped on the left inside of the visor flair with the manufacturers' mark and the size. The most prolific manufacturers being Eisenhuttenwerke Thale (marked ET), Sachsische Emaillirwerk A.G. Fulda (FS or EF), Vereinigte Deutsche Nickelwerke Schwerte (SE) and Quist Esslingen (Q). Differing manufacturing processes are apparent in the size of vent bushing, quality of finish and gauge of steel. For instance, Quist-produced helmets are nearly always a heavier gauge, and arguably, better finished than the others.

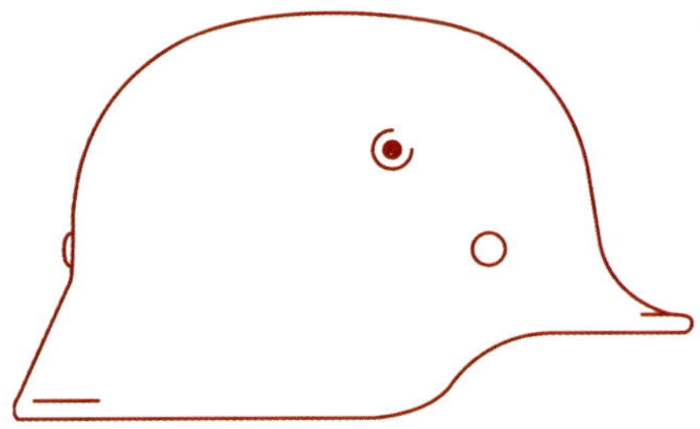

WORLD WAR II 1939-1945

It became apparent before the end of the first year of the war that the M35 needed modification. The paint-work was too smooth and shiny, especially when wet or at night. In March 1940, the production process was simplified, with the practice of inserting a bushing around the ventilation hole halted as soon as the existing supply was exhausted, after which it would be replaced with a simple embossed hole. A June 1940 directive required factories to replace the army and navy green finish with matte slate-grey and the airforce blue with a matte blue-grey. Troops in the field were to dull-down their helmets with camouflage. The directive also ordered that the national shield for the army, navy and airforce, and the state shield for the SS, would no longer be applied. Field troops were ordered to remove their shields. Nevertheless, many M35s retained both insignia. Manufacturers, pressed with the wartime urgencies of production, sometimes failed to make the changeover. And, almost needless to say, front-line troops often had more important matters to contend with.

A further simplification of the production process and therefore reduction in costs was ordered in July 1942. The inwardly rolled edge of the M35 and M40 was abandoned, with the M42 edge now simply outwardly crimped (the so-called "raw edge"). At about this time Eisenhuttenwerke Thale replaced its ET mark with "ckl" followed by the shell size, stamped inside the neck guard and Vereinigte Deutsche Nickelwerke Schwerte replaced its SE with "hkp".

In August 1943, production of all arms of service insignia – the army's silver eagle, the navy's gold eagle, the air force's eagle and the SS runes – ceased. As before, troops in the field were ordered to remove their helmet insignia, but, once again, not all did.

Towards the end of the war (1944/45), production of ventilation holes was stopped completely. Helmets of this late period are mostly smooth surfaced. Some bear insignia, some do not, although it is not certain whether the remaining insignia were applied in the factory or at the front, as sometimes happened when troops gained access to stores. This and many of these production simplifications were due to the progressive lack of materials and money as the war wore on.

Personal modifications by troops were commonplace throughout the war. Helmets were smeared with mud and clay. Texture was added by mixing paint with materials such as woodchips. Multi-coloured, painted camouflage was applied by hand and spray. Chicken-wire, straps and elasticised bands were added. A seemingly limitless variety of camouflage cloth covers and nets were improvised from odd pieces of material and attached by rubber-bands and home-made straps, as well as by the official drawstrings and metal clips.

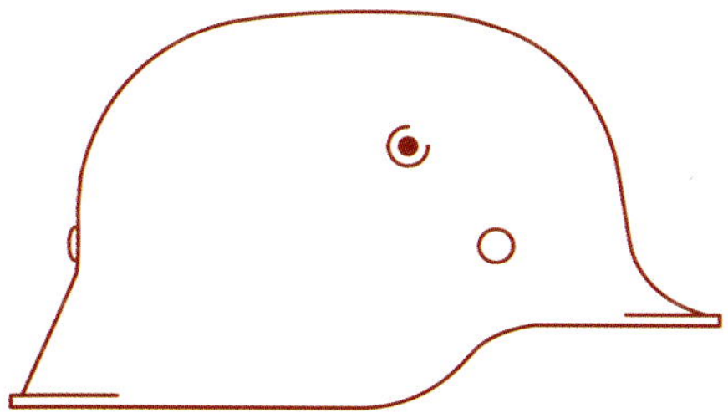

MILITARY POLICE

Germany's military police evolved from the myriad civilian units that operated throughout the country. Through a process of administration consolidation, they too ran the full gamut of equipment redesign. M16, 17 and 18 helmets, with modified or upgraded M31 liners, were worn initially, followed by the M35. Insignia proliferated from among the many civil police branches, but the discussion here will focus on the military police (Feldpolizei) which often fought alongside the regular army, Luftwaffe and Waffen-SS.

The two units issued with the combat-weight M35 were the Military Police and the 4th SS Police Division. Their M35s were initially produced alongside those for the main arms for service and in the same smooth apple-green. In July 1936 a new set of insignia was devised for all police units and fire brigades, throughout the country. On the left side of the helmet, under the ventilation hole was a silver eagle transfer with outstretched wings atop a swastika, set over a wreath of oakleaves in a black shield. On the right side was the swastika state insignia as used by the SS.

When the order of June 1940 was issued to "dull down" the shiny surface and remove the national and state shields, only the "dulling down" seems to have applied to the field police and the SS police division. The most likely reason was that, because of their continuing police duties, these units needed to remain distinguishable and apart from ordinary soldiers. Virtually all of the M42s that exist today with twin insignia belong to police units.

In 1944, the SS police division was eventually entirely incorporated into the armed SS, and their insignia conformed to the SS standard. These converted helmets have the SS runic shield on the right instead of the correct left side. It was easier for the new SS shield to be placed directly over the police shield without having to touch the state insignia.

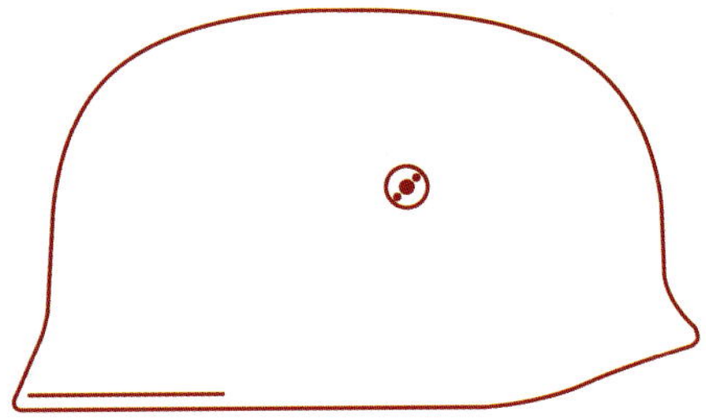

PARATROOP

Paratroops (fallschirmschützen) were the newest arm of the new airforce. The manpower for the first unit of battalion strength – established in January 1936 at Stendal was raised from volunteers of the Herman Göring Regiment. Among the range of specialist new equipment needed for these parachuting soldiers was a purpose-made helmet.

The first pattern – the Model 1936 (M36) – is very similar in shape to the army's M35, but with a much smaller flare around the shell brim to prevent snagging on parachute shroud lines. The chinstrap assembly comprises two leather two-piece straps which run from behind the head, then cross under the chin to reach the opposite temple. They are attached with carbine clips at each end and adjusted with basic prong buckles. These first helmets had two versions. One had four slots in the shell, in pairs at the front and back to take the carbine clips. The other has the four carbine clips attached directly to the liner band. There are two shell slots in this version, one on each side, to allow the cross-over chinstrap configuration for parachuting to be converted to a simpler configuration for combat. The slots are similar to those in the Vukanfiber M32 prototype. The M36 shell carries brushed ventilation holes similar to the M35's and is smooth-finished in the M35's apple-green.

The liner, which is very similar to the M31's, has up to nine leather tongues adjusted for fit by drawstrings. There are no ventilation holes, but thick felt cushioning has been added between the liner and shell. The liner band is attached to the shell with the M31's three split-cotter pin arrangement. Insignia follows the airforce style: the first-pattern droop-tail flying eagle on the left under the ventilation hole and the national tri-colour shield on the right.

In 1937, the army formed its own small paratroop force of one company, which was also issued with the airforce's M36 paratroop helmet. It was the same colour, but with the army eagle instead of the airforce version and with the same national tri-colour transfer on the right.

The paratroop M36 developed similar problems with manufacturing and use as the M35 steel helmet. The split cotter pins had a tendency to break under pressure. A complete redesign produced the Model 1938 (M38), which had a shell re-struck in the M36 shape but without brim slots. Four holes were drilled into the shell to attach the liner and chinstrap and provide ventilation. The liner assembly was heavily padded with light-yellow foam rubber. The old tongued arrangement was replaced by a two-piece section of leather with 12 large punched ventilation holes. This, together with a new chinstrap configuration, achieved a far more secure, compact helmet.

The chinstrap has a soft, chamois-backed strap attached to the right side of the shell by a vented nut and bolt through the shell and chinstrap. The other end passes through a sliding buckle which in turn attaches to a shorter strap secured to the liner on the other side of the helmet in the same way. An extra strap is sewn to each of the two side straps which cross over at the back of the head and are secured through the liner and shell by the nut and bolt system through the two holes at the rear. The finish retains the smooth apple-green.

Old M36 paratroop shells were converted to take the new liner by drilling two extra holes in the rear of the shell and utilising the existing ventilation holes on either side of the helmet. The replacement of the M36-style drawstring adjusting arrangement required the production of liners in various sizes. The army paratroop unit was expanded from company to battalion size and [they too] received the M38 upgrade, many of which now came with the airforce's second pattern eagle transfer.

In January 1939, the army's paratroop battalion was incorporated into the airforce's Paratroop Regiment 1 as the 2nd battalion and then, in June 1940, as the new Paratroop Regiment 3. Their helmets carried a mixture of army and airforce insignia.

As far as it is possible to say, all paratroop shells seem to have been produced by Eisenhuttenwerke Thale and marked accordingly inside the left brim above the rolled edge: eg. ET66, for shell size 66.

In June 1940, an order was issued that the shiny apple-green finish was to be replaced with a matte blue-grey and that the national shield was no longer to be applied. Once more, front-line troops were expected to remove theirs. And, again, this was not always done, to the undying gratitude of helmet collectors today.

In the autumn of 1940, a grey-green cloth cover was issued. This was attached to the helmet rim by six metal hooks. It has a darker strip of cloth sewn around the circumference of the helmet and a cloth cross-strap for inserting foliage for camouflage. Just before the airborne invasion of Crete – Operation Mercury in 1941 – a similar six-clip cover of splinter pattern camouflage material was issued. Later, cloth covers with several versions of "splinter" and "water tan" patterns were produced for attachment by clips and drawstring.

Wartime helmets differed from those of earlier manufacture in several ways. Pre- and early-war helmets had a large ventilation hole through the liner retaining nut and bolt with a small spanner adjustment hole on either side. One wartime version retains the central ventilation hole but incorporated a flat groove for screwdriver adjustment. A later type had no ventilation, only a screwdriver groove and was often made of aluminium. A very late wartime helmet utilises the M35-style split cotter pins, foregoing all ventilation. The foam padding between liner and shell was yellow in pre-war helmets and black for wartime, and later liners were produced in pigskin as well as the earlier sheepskin. The pre-war liners' ink-stamped makers' marks and sizings were in a serif typeface, while wartime ones tended to be in sans-serif. Chinstrap construction was also modified. Two versions followed the M38 sliding buckle pattern . The first was a sliding friction gate arrangement with snaps to secure the ends, the second later pattern was a sliding gripper friction buckle without snaps.

Helmet transfer insignia also changed for the airforce in general, with the production of a subdued second pattern grey-green eagle instead of the white variety. It is often assumed that this change did not appear until mid-war or even late-war, but M35 and M40 examples exist that carry the national shield and the subdued eagle.

Paratroop helmets, similarly to all of the many models and variations worn by front-line troops, have a vast array of camouflage: varieties of string nets (some captured from the enemy), painted colour and texture schemes sympathetic to surrounding flora and the season, and configurations of metal bands, straps and wire to disguise the helmet outline by inserting foliage. Some were factory-produced, some perhaps made at battalion or company level, but the vast majority were improvised by the individual soldier.

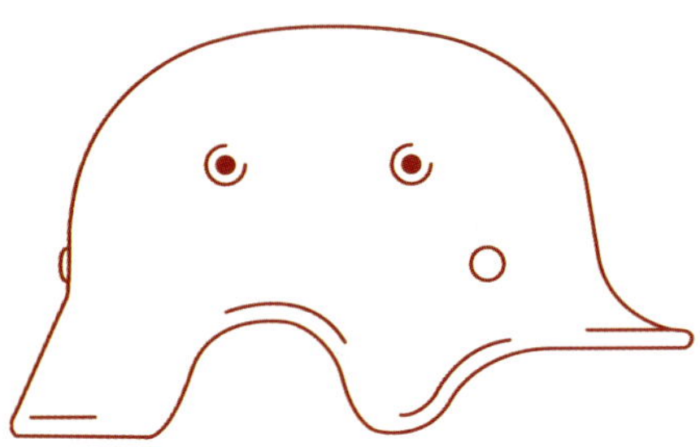

AIRCREW HELMETS 1939-1945

Aircrew flying helmets offered little or no protection for airforce pilots and crew under fire from anti-aircraft, cannon or machine-gun. M35s and M40s were sometimes worn over the flying helmet but were hampered in fit by the latter's earphones. Removing the steel helmets' M31 liners did not entirely solve the problem. Another idea tried was to remove the liners and beat out bulges in the shells so that they would slip over the earphones. These, however, are crudely constructed, probably by hand, with felt padding riveted into the shell and a leather Y-shaped chinstrap that joined under the chin with a prong or slip buckle.

In May 1941, the airforce authorised the production, by Siemens, of a purpose-made steel prototype – the SSK 90. It is made of overlapping steel plates covered in leather with cut-outs over the ears and a padded crash comb running from the front brim over the skull to just past the top. It was secured by a snap-fastened leather chinstrap, similar in construction to the paratroop M38's assembly. The SSK 90 prototype was tested but abandoned, possibly because its size was considered to be too large and/or its weight too heavy. Another prototype followed, also produced by Siemens, made of steel plates covered in leather with a chamois or cloth-lined interior. It was dome-shaped and had cut-outs over the earphones but smaller and simpler in design than in the SSK 90 but without the padded comb. Examples exist with the airforce breast eagle sewn onto the front. However, what was still required was a strong but light and comfortable helmet that could slip over the flying helmet and earphones and still obtain a decent fit.

The next development was a one-piece steel shell, again with cut-outs over the earphones, but with the rim rolled under inside and usually painted a matte grey-green but also in the colours of particular units, called staffel. The airforce flying-eagle transfer often appears on the front. Examples also exist with the insignia of the "flying Dutch clog" of the 26th Horst Wessel Squadron, painted below the eagle transfer. This implies that squadrons sometimes extended the use of "nose art" to their helmets.

The liner system consists of a metal band with a leather sweatband attached, with the top of the shell kept clear of the head by wide cloth cross-straps. Sets of twisted fibre cords form a V-shape, with the two top points attached to the liner either side of the cut-out and joining at the base of the V, from where a leather strap passes through a sliding buckle and back to a snap fixing. The configuration is very similar to the strap and fixing of the paratroop M38's. The liner was attached to the shell by six split cotter pins. It is ink-stamped with RB Nr. 0/0250/021 for the Heisler manufacturer, who also made the paratroop M38 liners, and with the shell and liner size. The Destroyer (Zerstörer) squadrons flying Messerschmitt 110s, 210s and 410s were most commonly the recipients of these late-war helmets.

Another end-of-war variation employed the M40-style shell with a rolled-under rim, double M40-type embossed ventilation holes on each side and an M31 liner and chinstrap attached with the usual M35, /M40/M42 configuration of three split cotter pins. Examples of a liner fixing similar to the French Model 1926/36 have come to light over the past few years and again, some with of a Y-shaped chinstrap, which must have improved the fit over the earphones and flying helmet.

All the wartime helmet changes and innovations involved attempts to improve both production methods and the end-product, but most, as the war wore on, reflected the over-riding need to cut costs in a nation beset by dwindling supplies and the increasing loss of manpower.

Further reading:
The History of the German Steel Helmet, Ludwig Baer
Von Stahlhelm zum Gefechtshelm, Ludwig Baer
Casques de Combats, Yves Plasseraud, Pierre-Paul Struye & Jean G. Daynas
Stahlhelm, Floyd R. Tubbs
Elmetti Di Tutto Il Mondo, Paolo Marzetti

MODEL 1916-18

■ M16 helmet shell with unknown unit insignia. It carries the Imperial Hohenzollern transfer (decal) and '271' painted over a red band with black top-stripe. '271' is not a regimental number, which would indicate that this could be a Freikorps unit helmet.

M16 helmet shell with Reichsmarine crossed anchor transfer. This insignia was authorised for wear from April 1924.

M16 helmet shell and liner, carrying the insignia of the Austrian Gardebataillon. This helmet is in the pre-Nazi form without any other insignia. The metal badge is attached to the shell by two nuts and bolts.

M16 helmet shell with M31 liner and M18 chinstrap. It carries a regulation Reichswehr shield, hand-painted on the left side. The shell has received at least one coat of regulation post-WWI apple green paint.

■ M16 helmet shell, re-painted apple green, with M31 liner and chinstrap. It carries the regulation Wehrmacht insignia: silver eagle transfer on the left side, with the tri-colour national shield transfer on the right.

M16 helmet shell with original WWI field grey paint and M31 liner and chinstrap. This helmet carries a transfer of the Weer Afdeelingen (Dutch NSB) on the right side. It appears that one of the M91 chinstrap lugs has either been removed or the helmet was never properly finished. This was probably a surplus unfinished WWI helmet re-issued to Germany's Fascist allies during WWII.

■ M16 helmet shell with M31 liner and chinstrap. It is unusual in that it carries only one Polizei/Feldpolizei transfer, possibly due to dwindling transfer stocks. (Collection of Bill Shea - The Ruptured Duck)

M16 helmet shell with M31 liner, no chinstrap. It carries a hand-painted Hitler Jugend insignia on the left side and hand-painted tri-colour national shield on the right.

M18 helmet shell and liner, no chinstrap, in its original late-WWI paint finish. This helmet carries the insignia configuration, unique to the Leibstandarte SS Adolf Hitler (LSSAH), of the earlier style SS runic transfer on the right and the tri-colour national shield on the left.

M18 helmet shell with M31 liner and pre-1931 chinstrap. Painted apple green, it carries the regulation Wehrmacht insignia of silver eagle transfer on the left side and tri-colour national shield on the right.

M18 helmet shell and M31 liner and chinstrap. This helmet is in its original late WWI finish with regulation twin Wehrmacht insignia. (Collection of Ellwood von Seibold)

M18 helmet shell with M31 liner and remnant of M31 chinstrap. This helmet is unusual in that it is sprayed in three colours used to camouflage military vehicles in the 1930s. It carries the remains of the regulation twin Wehrmacht insignia.

■ M18-style helmet shell, liner and chinstrap of post WWI manufacture, possibly made in the 1920s. The shell is thicker gauge than normal for M18 helmets and a feint crease across the top of the visor can be detected in certain lights, a feature which the author has noticed in several other post-WWI manufactured helmets. The liner and chinstrap, while generally conforming to the M18 configuration, also differs in several ways: the liner was made of thinner leather, which has survived wear and time better than its true M18 counterpart, and the chinstrap is of black leather, instead of the usual brown, and crudely finished. It carries the Bavarian coat of arms transfer placed over the ventilation lug on the left, attributed to the Bavarian Freikorps of 1919-1920.

■ M18 helmet with M31 liner and chinstrap. It carries the Wehrmacht eagle transfer on the left and the Lithuanian tri-colour national shield transfer on the right, and was used by Lithuanian volunteers to the German Army.

■ M18-style helmet shell with regulation M31 liner and chinstrap. The shell is painted dark Luftwaffe blue and displays the feint crease across the top of the visor. It carries the first pattern insignia of the Luftwaffe: the 'droop-tail' eagle transfer on the left side and the tri-colour national shield on the right.

M18 shell with M31 liner and chinstrap. The shell is painted a lighter blue than the previous helmet, and carries the second pattern Luftwaffe eagle transfer on the left with the tri-colour national shield on the right.

■ M18 helmet shell, liner and chinstrap. The shell is painted in the lighter version WWI field grey and carries the earlier style SS runic shield transfer on the right side with the swastika (state) shield transfer on the left. This configuration of helmet and insignia was worn by SS Verfügungstruppen (VT) and the LSSAH from August 1935.

M18 helmet shell, liner and chinstrap. This helmet is painted in the darker late-WWI field grey and carries the earlier style SS runic shield transfer on the right side with the swastika shield transfer on the left. This helmet could have been worn by either the SSVT or LSSAH units, and later after they had been amalgamated into the Waffen-SS.

■ M18-style helmet shell with M31 liner, no chinstrap. The shell is painted in an unusual blue-grey colour. It carries the insignia of the Polizei/Feldpolizei, with the Polizei eagle and wreath transfer on the left and the swastika shield transfer on the right. Another unusual feature is that the transfers have not been placed directly on the shell but on thin copper shields which are attached with pins throughout the helmet. (Collection of Bill Shea - The Ruptured Duck)

■ M18-style helmet shell with M27 liner and M31 chinstrap. This helmet is painted blue grey and carries the insignia of the Kriegsmarine: the gold eagle on the left side with the national tri-colour shield transfer on the right. The liner is permanently riveted to the shell with domed rivets, unlike the usual method of split cotter pins.

M18-style helmet shell with variant M27 liner and chinstrap. It is painted gloss black and carries early pre-Nazi era Assman-made metal shields. On the left side is the Bavarian state shield, and on the right the national tri-colour shield. The shields are attached to the shell with pins which pass through the shell, then bend back. While this Polizei helmet is strictly not from a combat unit, the helmet shell is high-quality and combat worthy. (Collection of Marcus Cotton)

M18 cut-out variant helmet shell, liner and chinstrap. Repainted apple green, it carries a half-completed tri-colour national shield on the left side, indicating that this was a pre-Nazi Reichswehr helmet. In all other respects this helmet is identical to the 1918 cut-out helmet of WWI.

■ M18 cut-out variant shell with M31 liner and variant chinstrap. This helmet has also been repainted apple green and carries twin Wehrmacht transfers.

M18 cut-out variant shell with M31 liner and chinstrap. It is painted apple green and carries only one transfer: the Wehrmacht silver eagle on the left side. Although of early manufacture, it was probably pressed into service from surplus WWI stores in about 1940, and conforms to the 1940 regulation prohibiting the further use of the national tri-colour shield. (Collection of Bill Shea - The Ruptured Duck)

M18 'droop-nose' variant helmet shell with M27 liner and chinstrap. It is painted in regulation WWI field grey and carries no insignia.

M17 Austrian helmet shell with M31 liner and variant chinstrap. It is painted regulation WWI field grey and carries the twin Wehrmacht insignia: the silver eagle transfer on the left side, with the national tri-colour shield transfer on the right.

■ M17 Austrian helmet shell with M31 liner and chinstrap. It is painted satin black and carries twin insignia for the SS: earlier pattern SS runic shield transfer on the right side, and swastika state shield transfer on the left. (Collection of Ellwood von Seibold)

■ M17 Austrian helmet shell with M31 liner and chinstrap. It is painted apple green and carries the twin insignia of the SS. Unusually, the earlier pattern runic shield transfer on the right side has been reversed, possibly due to a manufacturing error. (Collection of Bill Shea - The Ruptured Duck)

MODEL 1932

M32 Vulkanfiber prototype No.2 helmet, liner and chinstrap. This series of helmets was proposed as a replacement for the M16/M18 style helmets of WWI. They were of lightweight composite construction and braced around the edge with a metal band. Painted field grey, this helmet carries the twin insignia of the Wehrmacht. One would be forgiven in thinking this was a parade, and not a combat, helmet. However, it was seriously considered for combat use, sacrificing protection for comfort and visibility. (Collection of Brian Maederer - Military Collectibles)

It was originally intended that an Assmann made metal tri-colour national shield would be used with the pins for attachment passing through the ready-made, grommeted holes below the ventilation screen (clearly visible in this photograph).

Rear view.

The improved liner system for the M32.

M32 Vulkanfiber prototype No.3 helmet. It differs from prototype No.2 in that it has two oblong, lozenge-shaped cut-outs on each side of the neck apron above the ears. These were possibly added either for ease of transport (strapped to a back pack), or to allow the wearer to vary the way the chinstrap was fastened.

Front and rear views.

Details of helmet, liner and chinstrap.

MODEL 1935

M35 helmet shell with M31 liner and chinstrap. This represents the archetypal configuration of the German helmet of WWII. Painted satin apple green, the new cut-down shell carries the twin insignia of the Wehrmacht. (Collection of Ellwood von Seibold)

M35 helmet shell with M31 liner and partial chinstrap. It is also painted satin apple green, somewhat darker than the previous example, and carries the twin insignia of the Wehrmacht. (Collection of Brian Maederer - Military Collectibles)

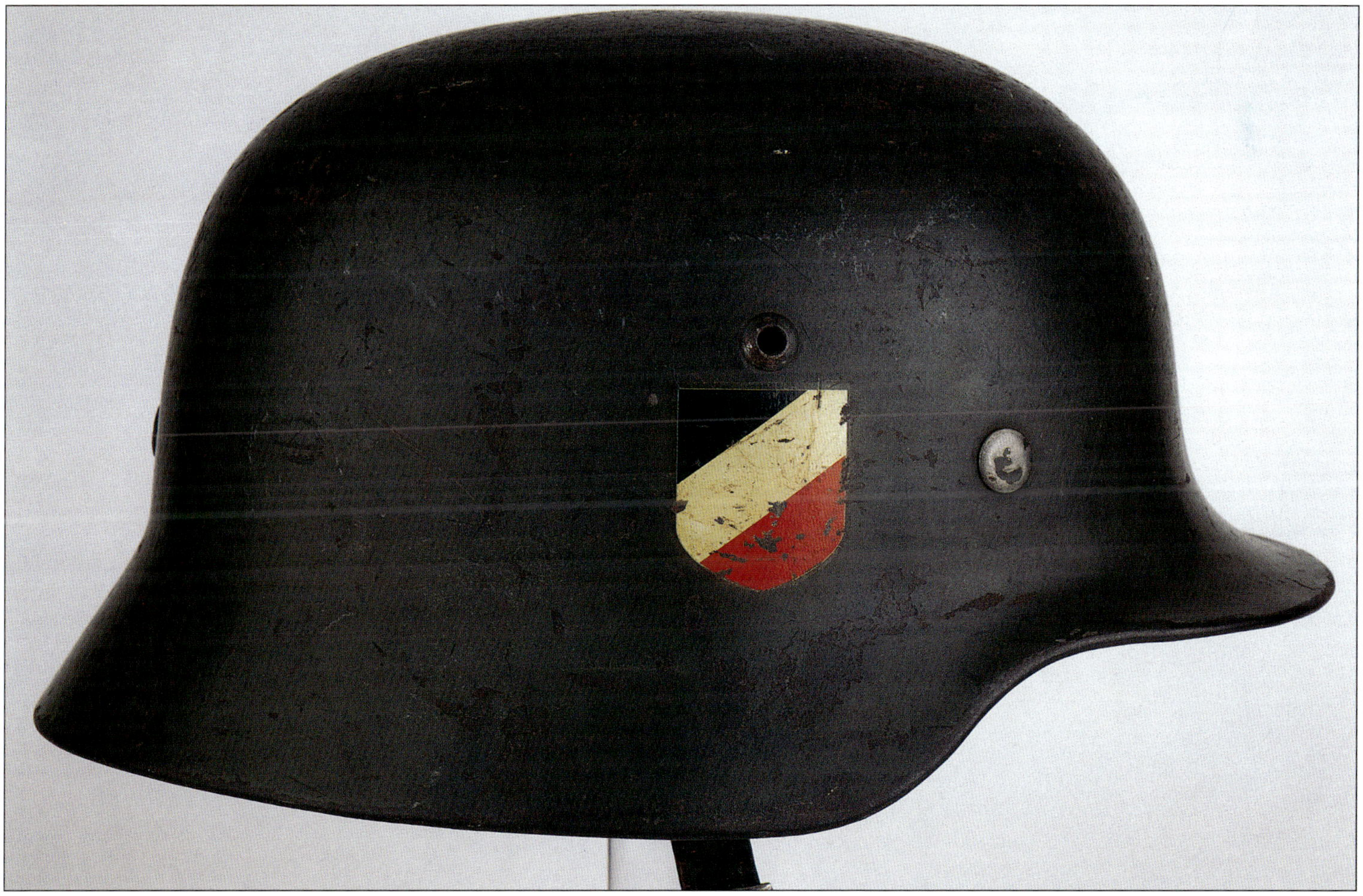

M35 helmet shell with M31 liner. It is painted gloss apple green and carries the twin insignia of the Wehrmacht. Of note is the size and shape of the bushing around the ventilation holes compared with the previous helmet.

M35 helmet shell with M31 liner and chinstrap. It is painted satin apple green and carries the twin insignia of the Wehrmacht.

M35 helmet shell with M31 liner and chinstrap. This helmet is painted a satin apple green colour and carries the twin insignia of the Kriegsmarine: the gold eagle transfer on the left side, with the national tri-colour shield transfer on the right. (Collection of Brian Maederer - Military Collectibles)

M35 helmet shell with M31 liner and chinstrap. This helmet is painted a gloss blue colour and carries the twin insignia of the Luftwaffe: the early 1936 'droop tail' white eagle transfer on the left side, with the national tri-colour shield transfer on the right. (Collection of Ellwood von Seibold)

M35 helmet shell with M31 liner and chinstrap. This helmet is painted a satin light blue colour and carries the twin insignia of the Luftwaffe: the later version of the white eagle transfer on the left side, with the national tri-colour shield transfer on the right.

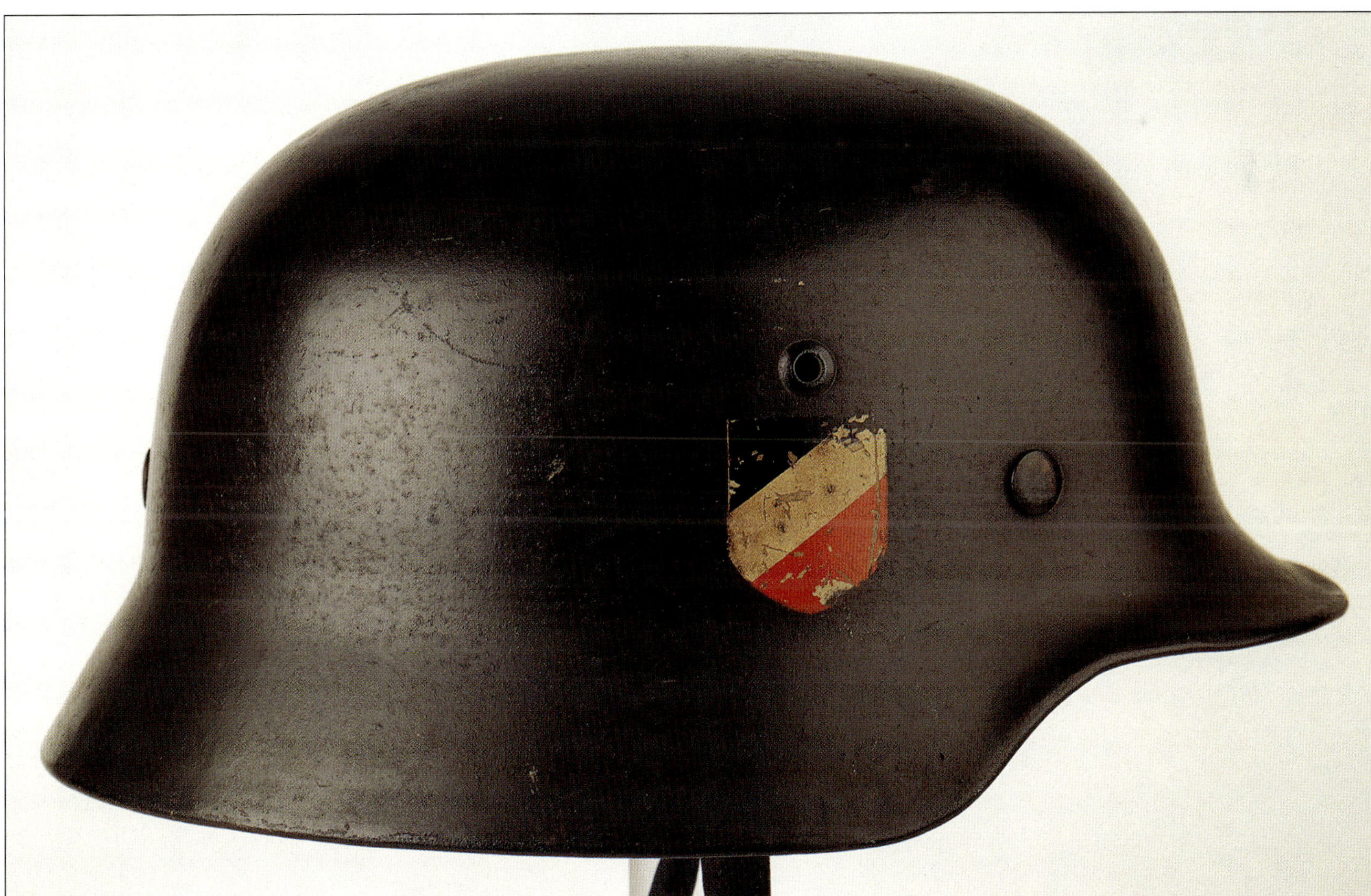

M35 helmet shell with M31 liner and chinstrap. This helmet is painted a satin apple green colour and carries the twin insignia of the SS: the early SS runic shield transfer on the left side, with the swastika state shield transfer on the right.

M35 helmet shell with M31 liner and chinstrap. This helmet is painted a gloss SA brown colour and carries the twin insignia of the SA-Standarte Feldherrnhalle: on the left side is the Feldherrnhalle shield transfer, and on the right is a swastika shield. This helmet was worn by members of the detachment which formed the guard for the Feldherrnhalle monument in Munich, in which the bodies of those killed during the 1923 Putsch were placed. (Collection of Brian Maederer - Military Collectibles)

M35 helmet shell with M31 liner and chinstrap. This helmet, originally Luftwaffe blue, has been over-painted in a satin SA brown. It carries twin insignia: on the left side the later pattern Luftwaffe white eagle transfer, and on the right the Feldherrnhalle shield transfer. In 1938 the SA Standarte Feldherrnhalle was redesignated as a regiment belonging to the Luftwaffe, and the insignia was changed accordingly as shown above.

M35 helmet shell with M31 liner and chinstrap. This helmet is painted a satin apple green and carries the twin insignia of the SS in the normal configuration of early SS runic shield transfer on the right and swastika state shield transfer on the left. In this particular case however, we see that the runic shield has been placed over the original tri-color national shield transfer, and the swastika state shield over the silver eagle transfer of the Wehrmacht. Clearly, this was originally an army twin transfer helmet re-used by the SS. (Collection of Brian Maederer - Military Collectibles)

■ M35 helmet shell with M31 liner and chinstrap. This helmet is painted a gloss black and carries the twin insignia of the SS: early SS runic shield on the right and swastika state shield on the left. The chinstrap is of SS manufacture and is stamped 'SSBW' into the leather.

M35 helmet shell with M31 liner and chinstrap. This helmet is painted a satin Luftwaffe blue-grey and carries the twin insignia of the Luftwaffe: early 'droop tail' eagle transfer on the left, and national tri-color shield transfer on the right. Around the helmet is a 1939 dated leather carrying strap for attachment to pack or belt.

M35 helmet shell with M31 liner and chinstrap. This helmet was originally painted a satin apple green, but has been overpainted with a matte, slightly textured, field-grey colour which would indicate its use during the early stages of WWII. It carries the twin insignia of the Wehrmacht: the silver eagle transfer on the left side, with the national tri-colour shield transfer on the right.

M35 helmet shell with M31 liner and chinstrap. This helmet is painted a matte field-grey colour and carries the eagle transfer of the Wehrmacht, not only on the left side, but also on the front. It has been suggested that this could have been for use by Hitler Youth units during the latter part of WWII, and the fact that this is a small size helmet would seem to support this, but there is no evidence that this is in fact the case.

M35 helmet shell with M31 liner and chinstrap. This helmet was over-painted a matte field-grey colour. It was originally a twin transfer army helmet in a smooth apple green colour. The twin transfers are just visible through the flaking paint.

M35 helmet shell with M31 liner and chinstrap. This helmet was originally without insignia and painted a matte field-grey. It has been over-painted in white with a red cross emblem painted on to the front. This helmet would have been worn by Sanitäts (medics).

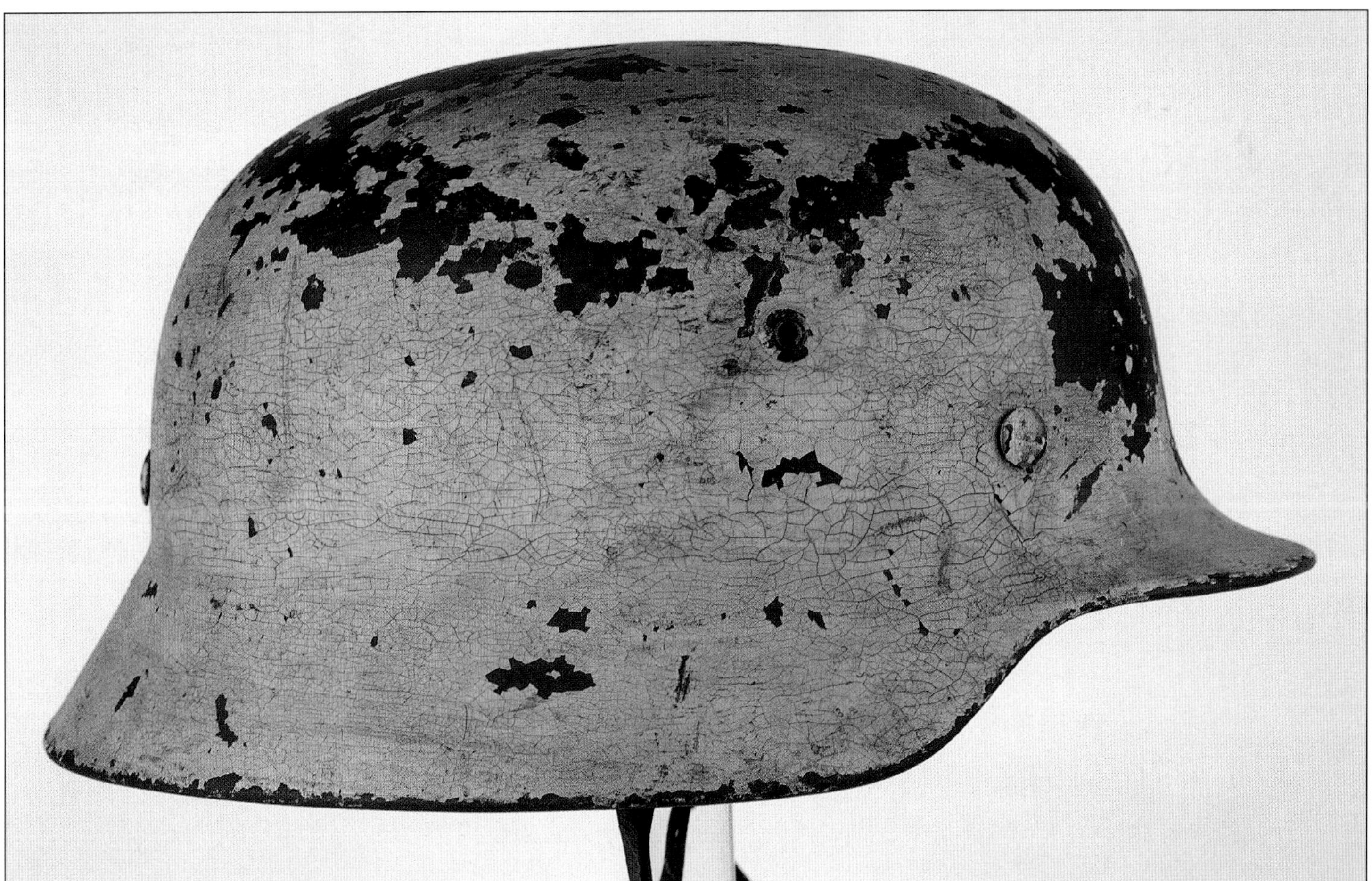

M35 helmet shell with M31 liner and chinstrap. This helmet is a double transfer army helmet, originally painted in satin apple green but has been over-painted in white for use on the eastern front. This helmet was brought back from Moscow by the author.

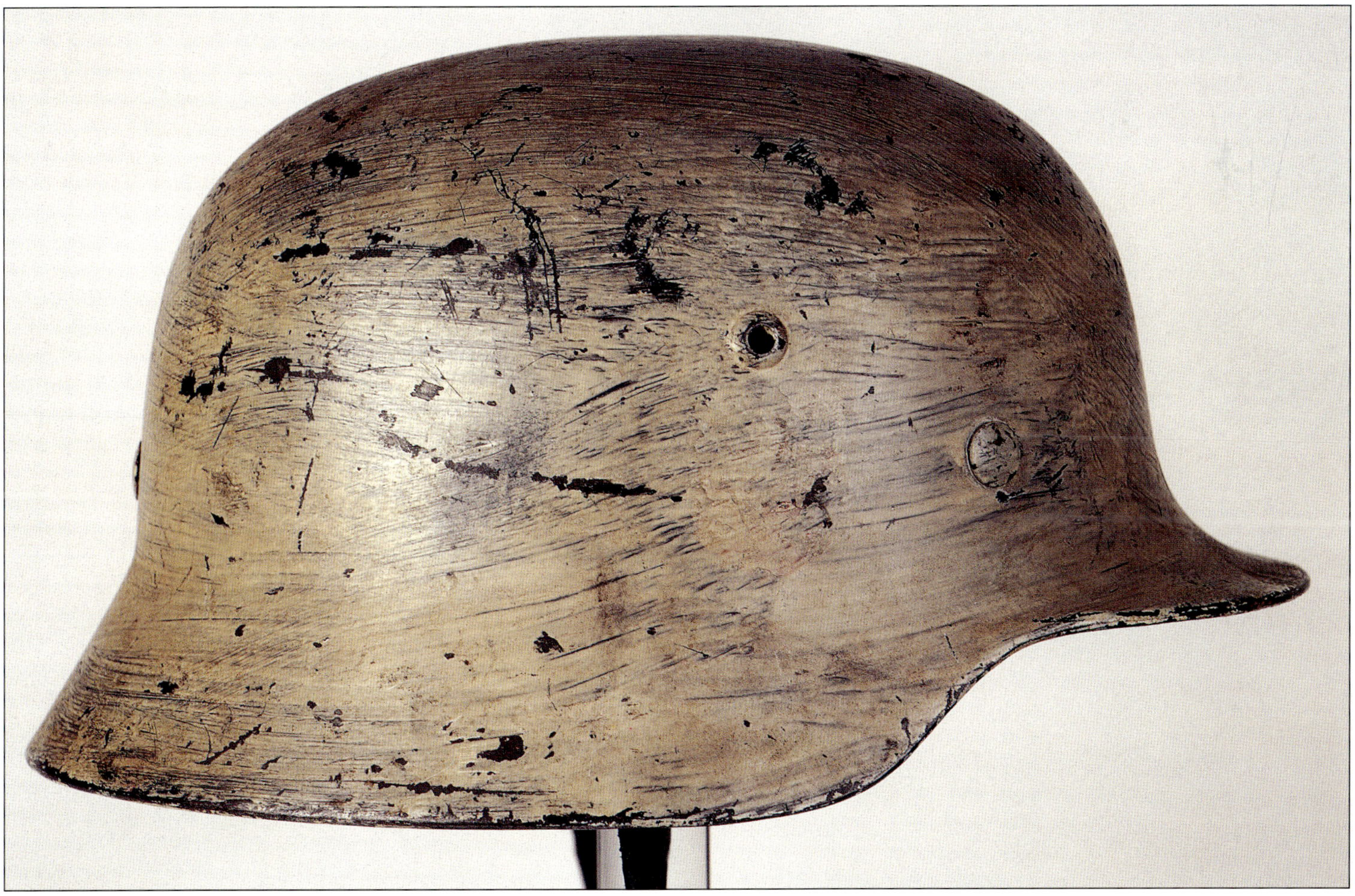

■ M35 helmet shell with M31 liner and chinstrap. This helmet carries the single insignia of the army in its 1940 configuration: silver eagle transfer on the left side. It has been over-painted in white, except around the eagle transfer. This helmet was for use in winter and probably came from the western front as it was found in France many years ago.

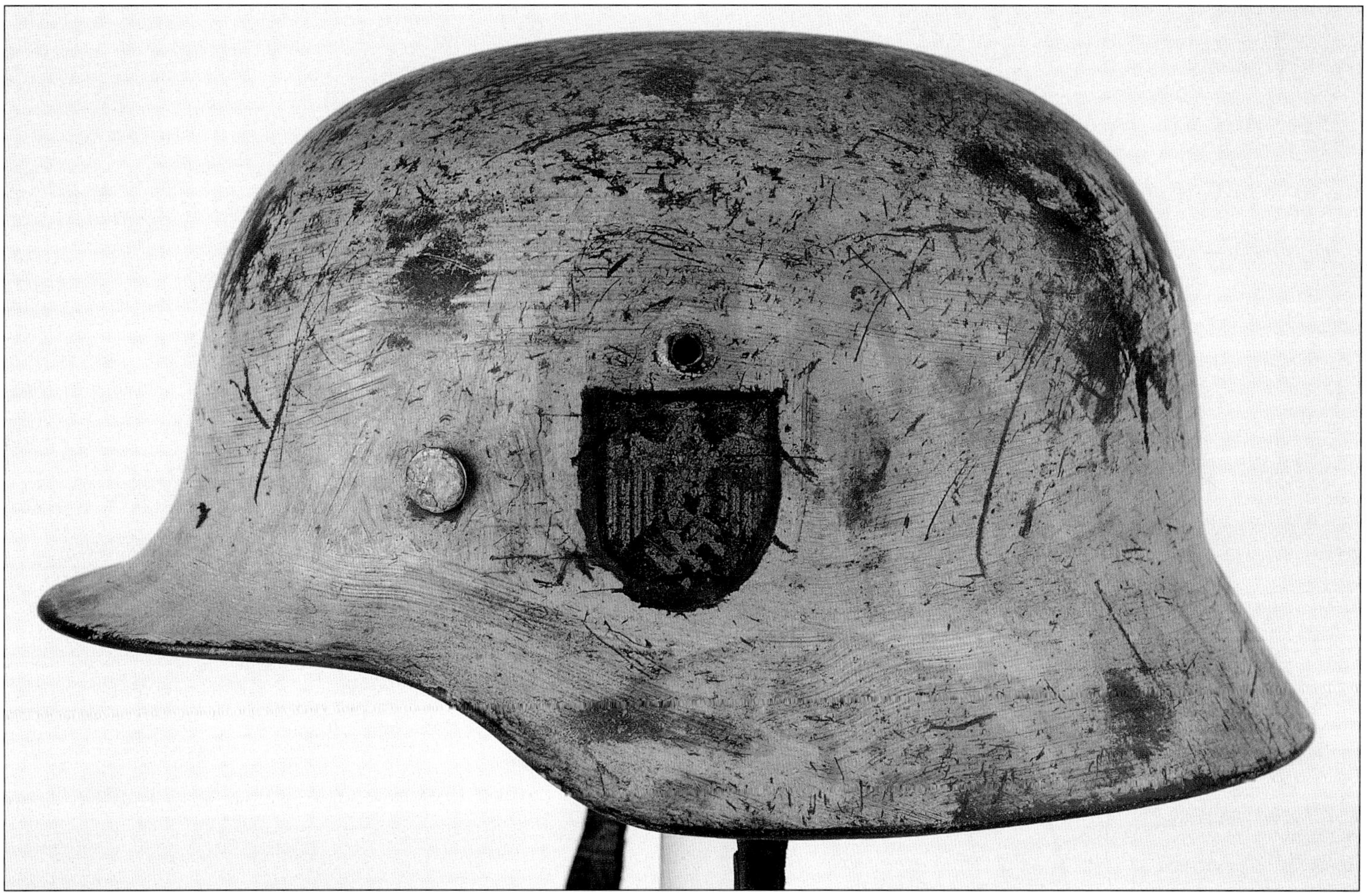

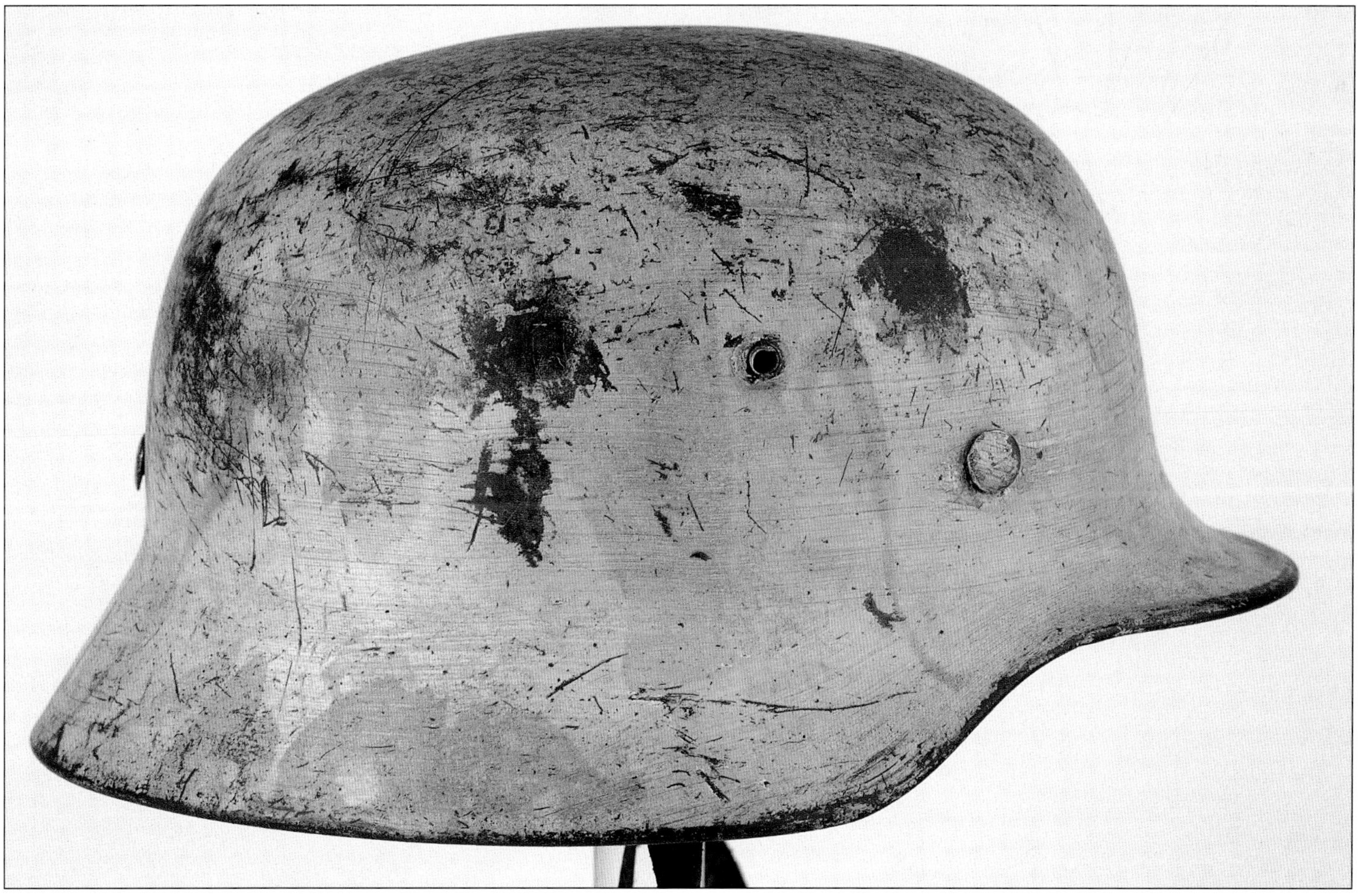

■ M35 helmet shell with M31 liner and chinstrap. This helmet carries the single insignia of the army in its 1940 configuration: silver eagle transfer on the left side. It has been over-painted in a sand colour except around the eagle transfer. It's impossible to say whether this camouflage was for North Africa, the Caucasus, or Italy.

M35 helmet shell with M31 liner and chinstrap. This helmet carries the twin insignia of the Luftwaffe but with the early 'droop tail' eagle transfer. It has been over-painted in sand colour right up to, but not over, the insignia. Inside the helmet shell was found a cut open Luftwaffe flight glove, presumably to help shield the wearer from intense heat. This particular 'pinkish' sand colour is often attributed to North Africa.

■ M35 helmet shell with M31 liner and chinstrap. This helmet carries the twin insignia of the Luftwaffe but with the later eagle transfer and has been over-painted in sand colour around the insignia. The liner has a printed label sewn in which reads, 'Huber I 2./Flak-Regt 5.'

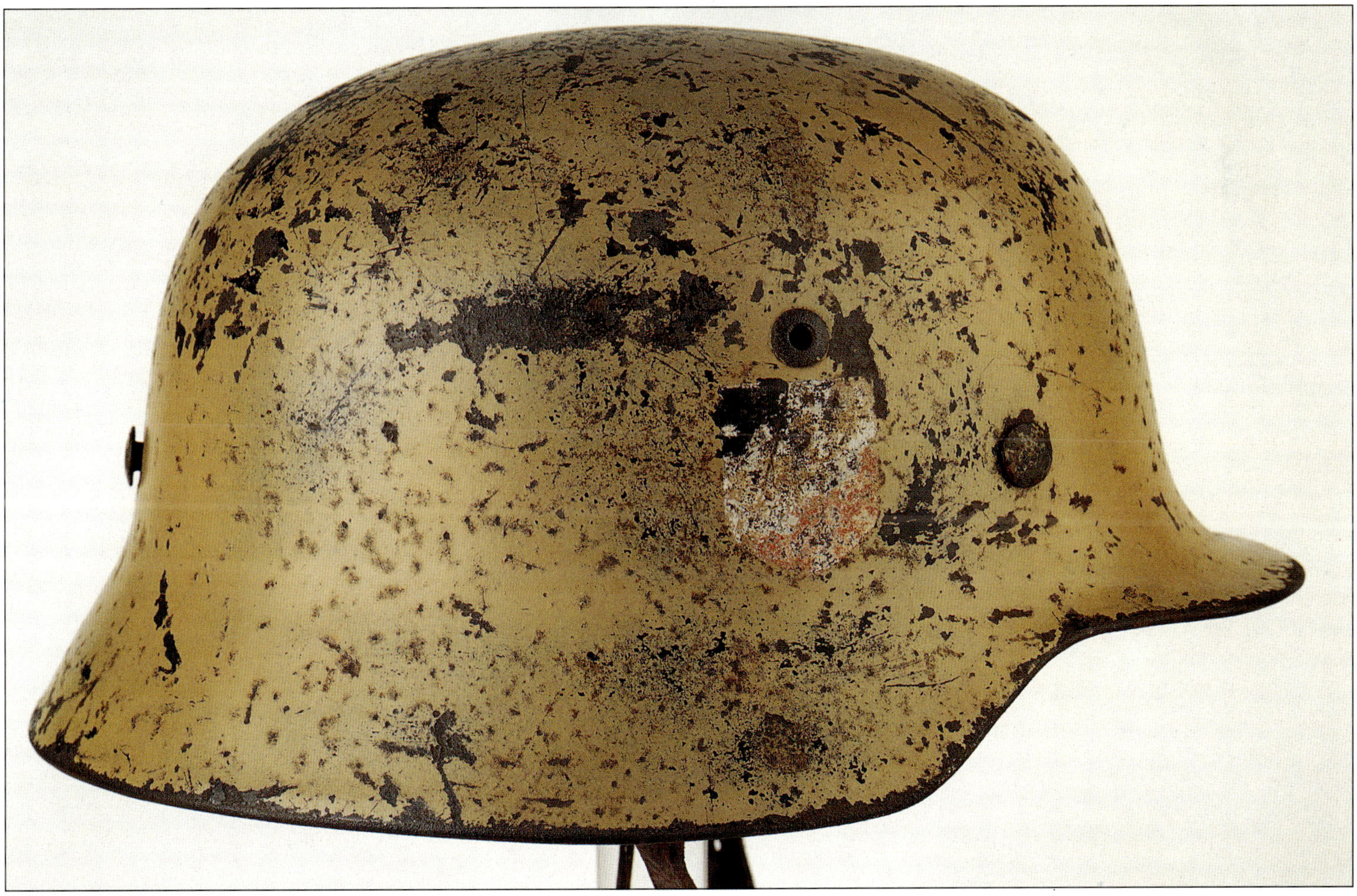

M35 helmet shell with M31 liner and chinstrap. This helmet carries the twin insignia of the Luftwaffe but with the later eagle transfer and has been over-painted in sand colour around the insignia. (Collection Bill Shea - The Ruptured Duck)

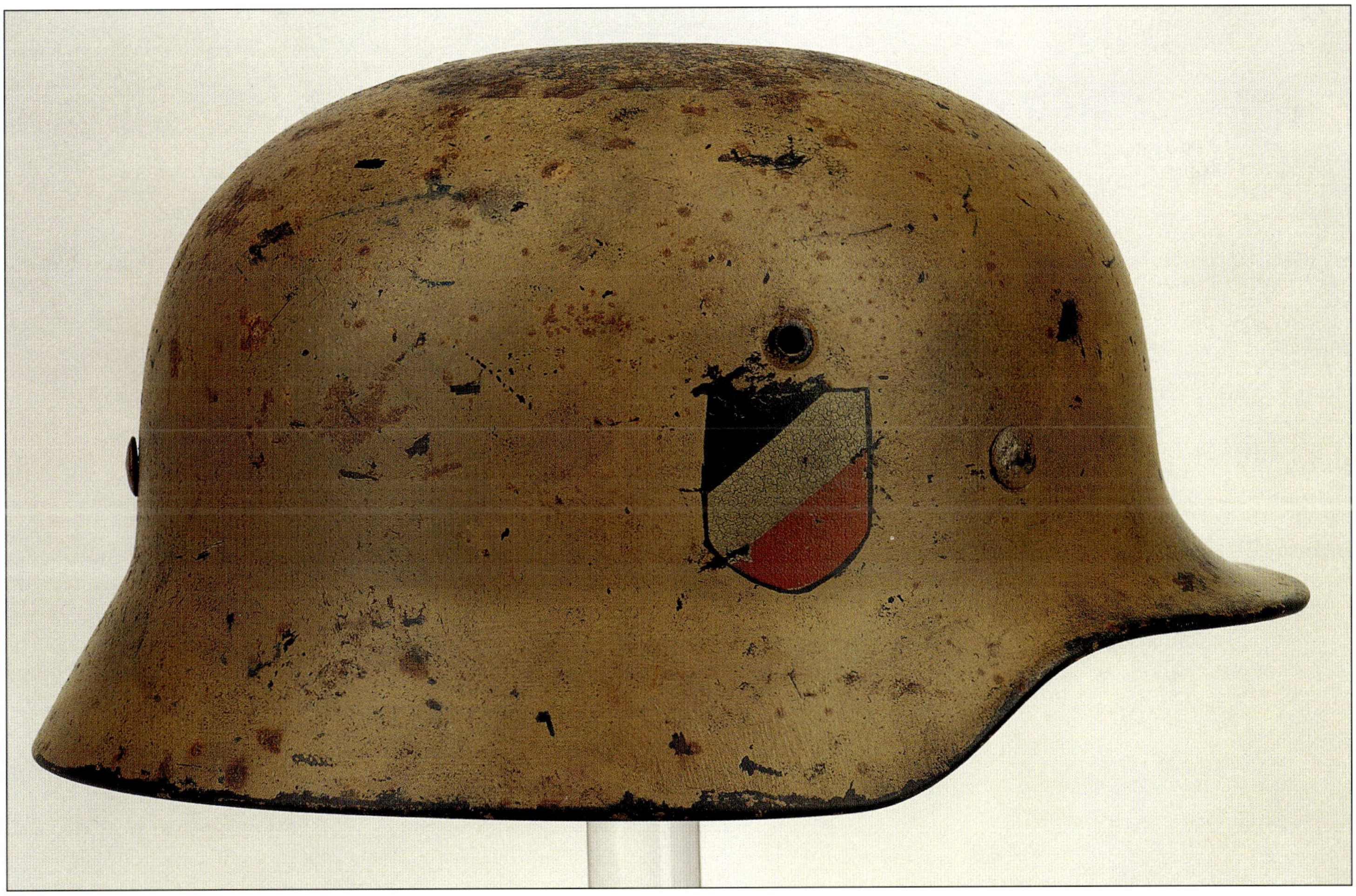

M35 helmet shell with M31 liner and chinstrap. This helmet was originally painted a matte field-grey and carries the twin insignia of the SS with the earlier runic shield on the right. It has had two subsequent over-paints: the first being white, painted around the insignia, the second a matte field-grey. This helmet displays the common practise of repainting after the spring thaw.

M35 helmet shell with M31 liner and chinstrap. This helmet is painted a satin apple green/grey and carries the twin insignia of the SS with the earlier runic shield on the right. (Collection of Bill Shea - The Ruptured Duck)

M35 helmet shell with M31 liner and chinstrap. This helmet was originally painted a satin apple green and carries the twin insignia of the SS with the earlier SS runic shield on the right. It has been camouflaged in sand and brown colours avoiding the insignia.

M35 helmet shell with M31 liner and chinstrap. This helmet was originally painted a matte field-grey and has been over-painted in colours mimicking the splinter pattern camouflage. It carries the twin transfers of the Feldpolizei: the police eagle transfer on the left, and swastika state shield transfer on the right. This is a much used helmet, it has three different owner's names inside.

M35 helmet shell with M31 liner and chinstrap. This helmet was originally painted a satin apple green and has been over-painted in patches of brown, sand, and khaki. It carries the twin insignia of the Feldpolizei.

M35 helmet shell with M31 liner and chinstrap. This helmet has been over-painted in a light sand textured colour. If any insignia was originally in place it is now entirely hidden.

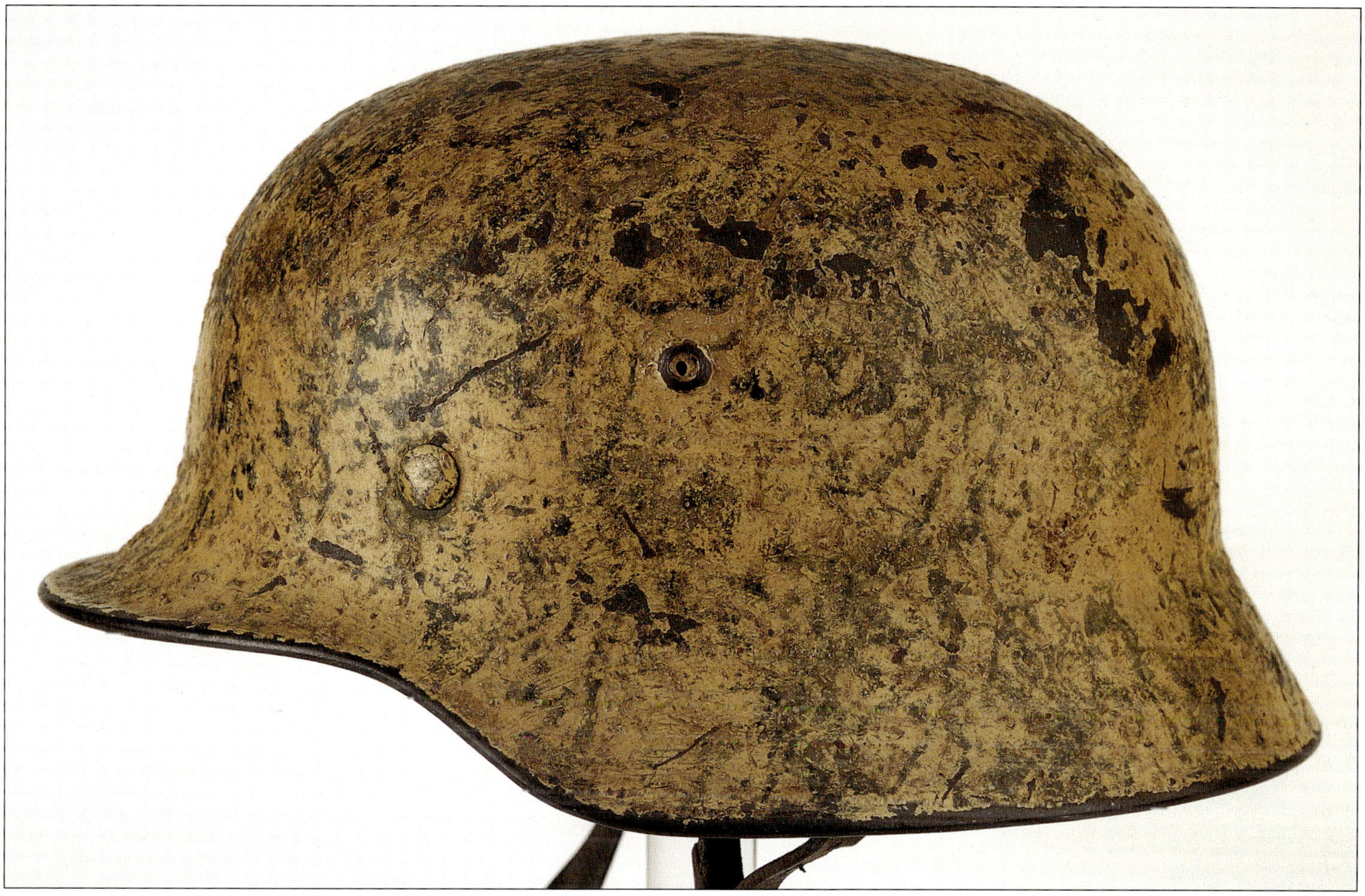

M35 helmet shell with M31 liner and chinstrap. This helmet has been carelessly brushed with white paint. It was originally painted a matte field-grey, and had its single army insignia removed from the left side as per the order of August 1943, requiring the complete removal of any army service insignia.

M35 helmet shell with M31 liner and chinstrap. This helmet was originally a matte Luftwaffe blue-grey without insignia and has been repainted in brown, sand, and green colours.

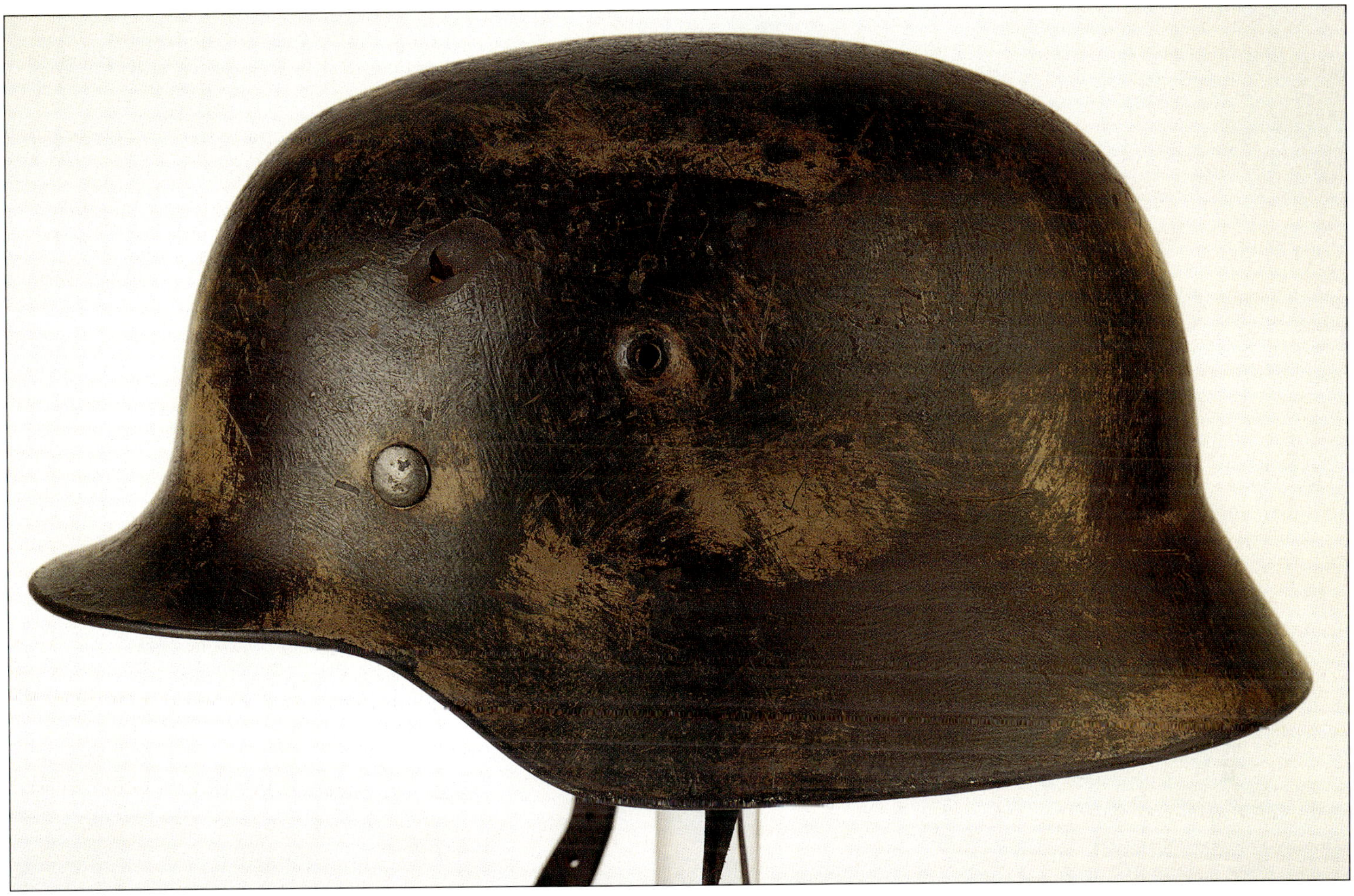

M35 helmet shell with M31 liner and chinstrap. This helmet seems to have originally been a twin insignia Wehrmacht helmet painted in satin apple green. It has at least two layers of subsequent paint applied: the first a white, and the second patches of green, brown, and sand very similar in style to some German WWI helmet camouflage.

Inset, opposite above: Carved into the camouflage is a heart shape with the name 'Inga' underneath, presumably the owner's wife or sweetheart.

M35 helmet shell with M31 liner and chinstrap. This helmet has been coated in a heavy 'Zimmerit' like paste and over-sprayed in brown, green, and sand. no insignia is visible.

M35 helmet shell with M31 liner and chinstrap. This helmet is painted a matte field-grey and carries a single Wehrmacht insignia of a silver eagle transfer on the left side. Over the top has been placed a cage of heavy gauge 'chicken wire' for the attachment of foliage. This helmet was found in the British Channel Islands, off the coast of France, which was the only part of the UK to have been occupied by the German Army during WWII.

M35 helmet shell with M31 liner and chinstrap. This helmet is painted a matte field-grey and carries the single insignia of the Kriegsmarine: a gold eagle transfer on the left side. Wire has been added for the attachment of foliage for camouflage.

■ M35 helmet shell with M31 liner and chinstrap. This helmet has been painted a matte field-grey and carries the remains of the late pattern Luftwaffe eagle transfer on the left side, and the complete national tri-color shield transfer on the right. Over the top as been placed riveted, creased steel strips through which string has been passed to facilitate the attachment of foliage.

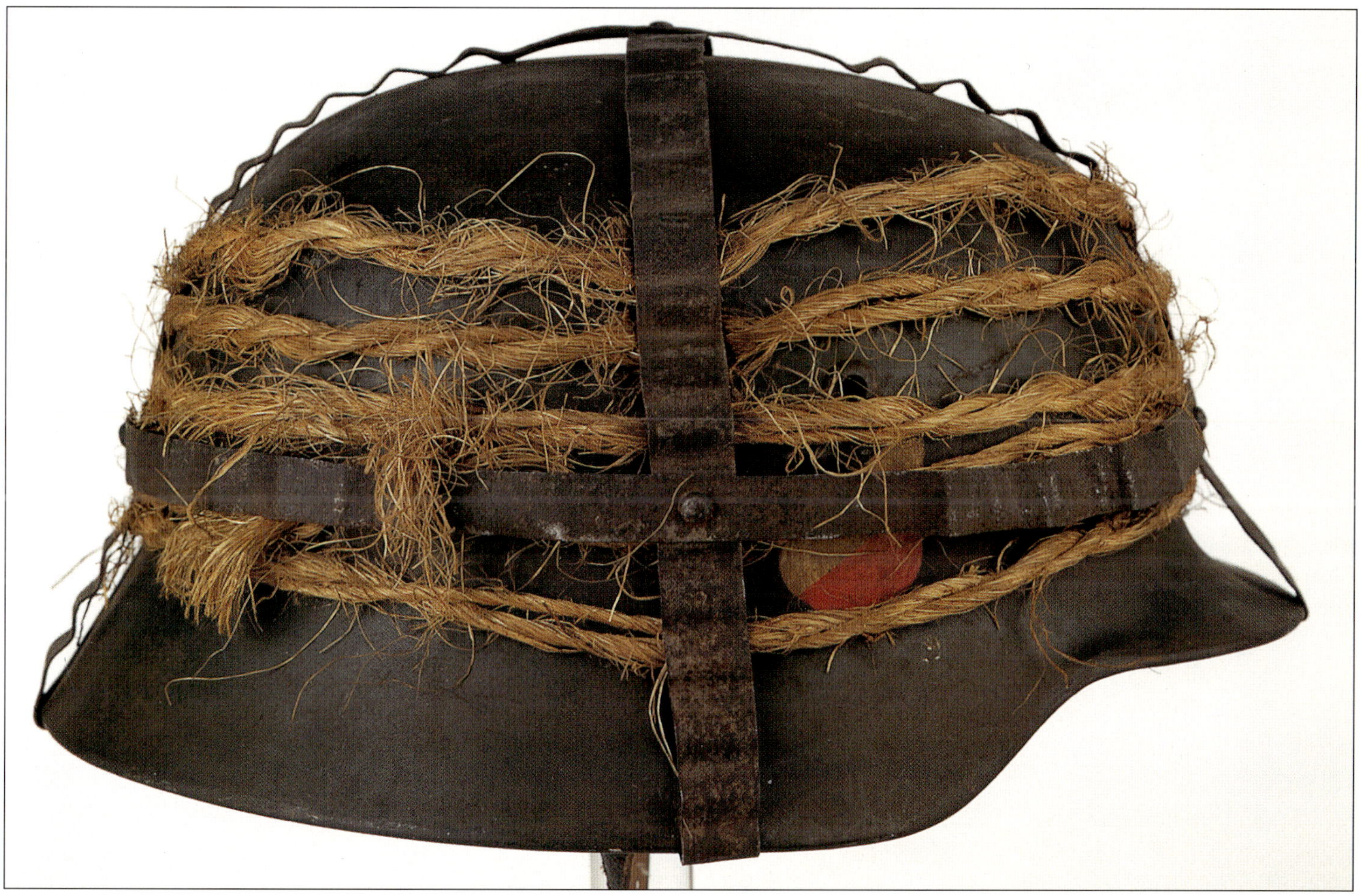

M35 helmet shell with M31 liner and chinstrap. This helmet has been painted a matte field-grey and carries no insignia. A bread bag strap has been added for the attachment of camouflage.

M35 helmet shell with M31 liner and chinstrap. This helmet was originally painted a matte field-grey and has been over-painted with patches of sand colour. A purpose-made strap with clips has been added for the attachment of camouflage. (Collection of Bill Shea - The Ruptured Duck)

MODEL 1940

M40 helmet shell with M31 liner and chinstrap. This helmet has been painted a matte field grey and has the gold Kriegsmarine eagle transfer insignia on the left side. (Collection of Bill Shea - The Ruptured Duck)

■ M40 helmet shell with M31 liner and chinstrap. This helmet has been painted a matte field grey and has the late pattern Luftwaffe eagle transfer insignia on the left side.

■ M40 helmet shell with M31 liner and chinstrap. This helmet has been painted a matte field grey and has the earlier SS runic shield transfer insignia on the right side. As can be seen from the photographs, this helmet has never been worn. The matte paint finish is extraordinarily good and reflects no light.

M40 helmet shell with M31 liner and chinstrap. This helmet has been painted a matte field grey and has the earlier SS runic shield transfer insignia on the right side. Unlike the previous helmet this has seen some wear, but it demonstrates quite well, how the paint finish deteriorates with wear over time. This helmet was brought home by T/5 William R. Batten, Battery C, 108th Anti-aircraft Artillery Gun Battalion. It is named to SS-Rottenführer Manthey in ink on the liner. (Collection of Barry Smith)

M40 helmet shell with M31 liner and chinstrap. This helmet has been painted a matte field grey and carries the twin insignia of the Feldpolizei.

■ M40 helmet shell with M31 liner and chinstrap. This helmet has been painted a matte field grey and has the Wehrmacht silver eagle transfer on the left side and the Spanish Blue Division shield transfer on the right.

M40 helmet shell with M31 liner and chinstrap. This helmet has been painted a matte field grey and has a later pattern SS runic shield transfer on the left side with a French tri-colour shield transfer on the right. It was worn by members of the SS Charlemagne Division.

M40 helmet shell with M31 liner and chinstrap. This helmet has been painted a matte field grey and has a late pattern SS runic shield transfer on the right side. It is not clear why the transfer is on the right side instead of the usual left.

M40 helmet shell with M31 liner and chinstrap. This helmet has been painted a matte field grey and carries the earlier pattern SS runic shield on both sides. It has been suggested that this was for use by foreign volunteers of the Waffen-SS, but period photographs show that configuration of insignia being worn by members of a German Waffen-SS Division in 1940. These photographs can be seen in the second volume of this book.

M40 helmet shell with M31 liner and chinstrap. This helmet has been camouflaged over a base sand colour, with patches of green and brown. No insignia is visible.

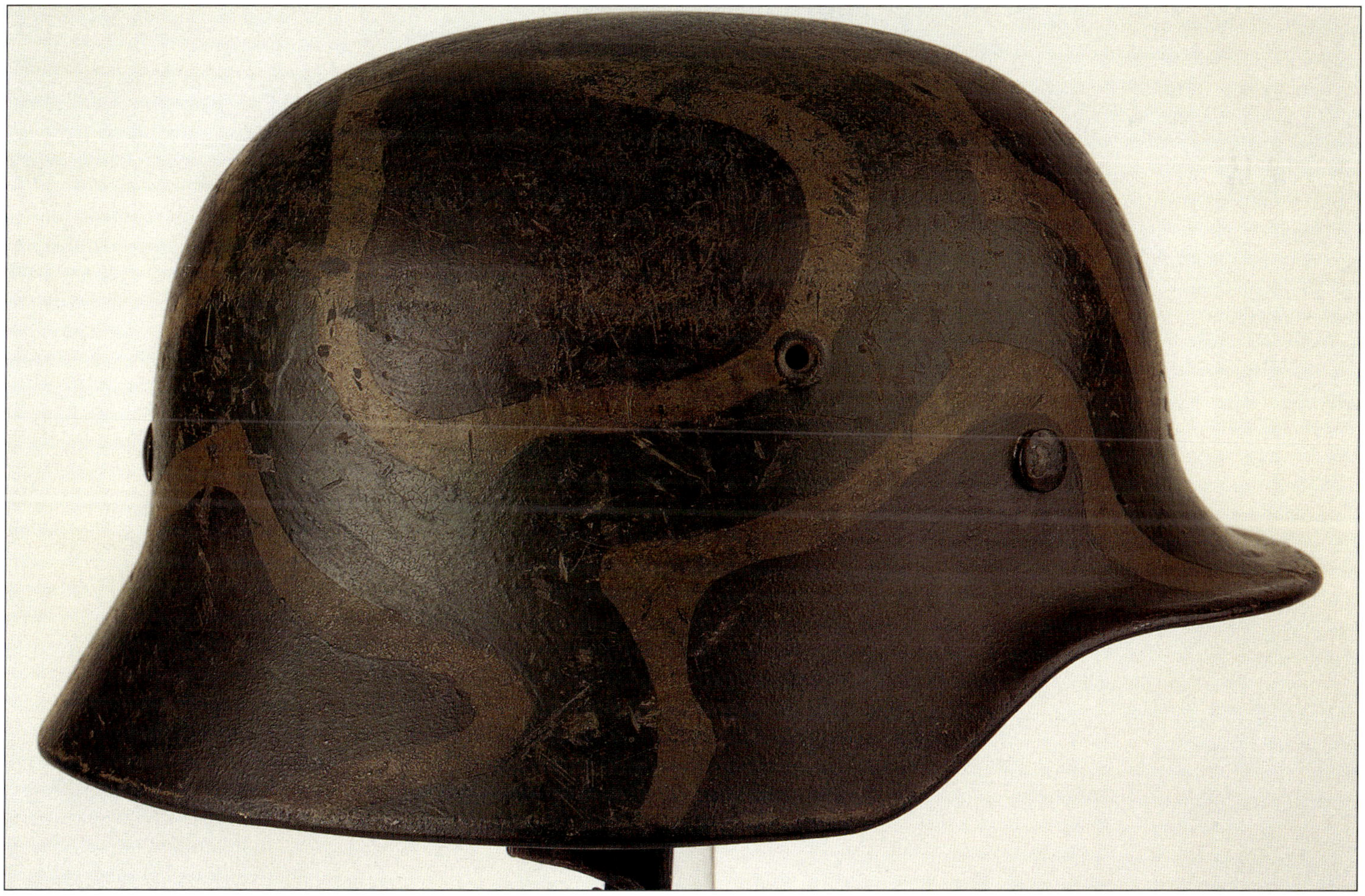

■ M40 helmet shell with M31 liner and chinstrap. This helmet was originally painted a matte field grey and has been overpainted in white with no insignia visible. It was used during the Russian campaign and was found in the Leningrad area.

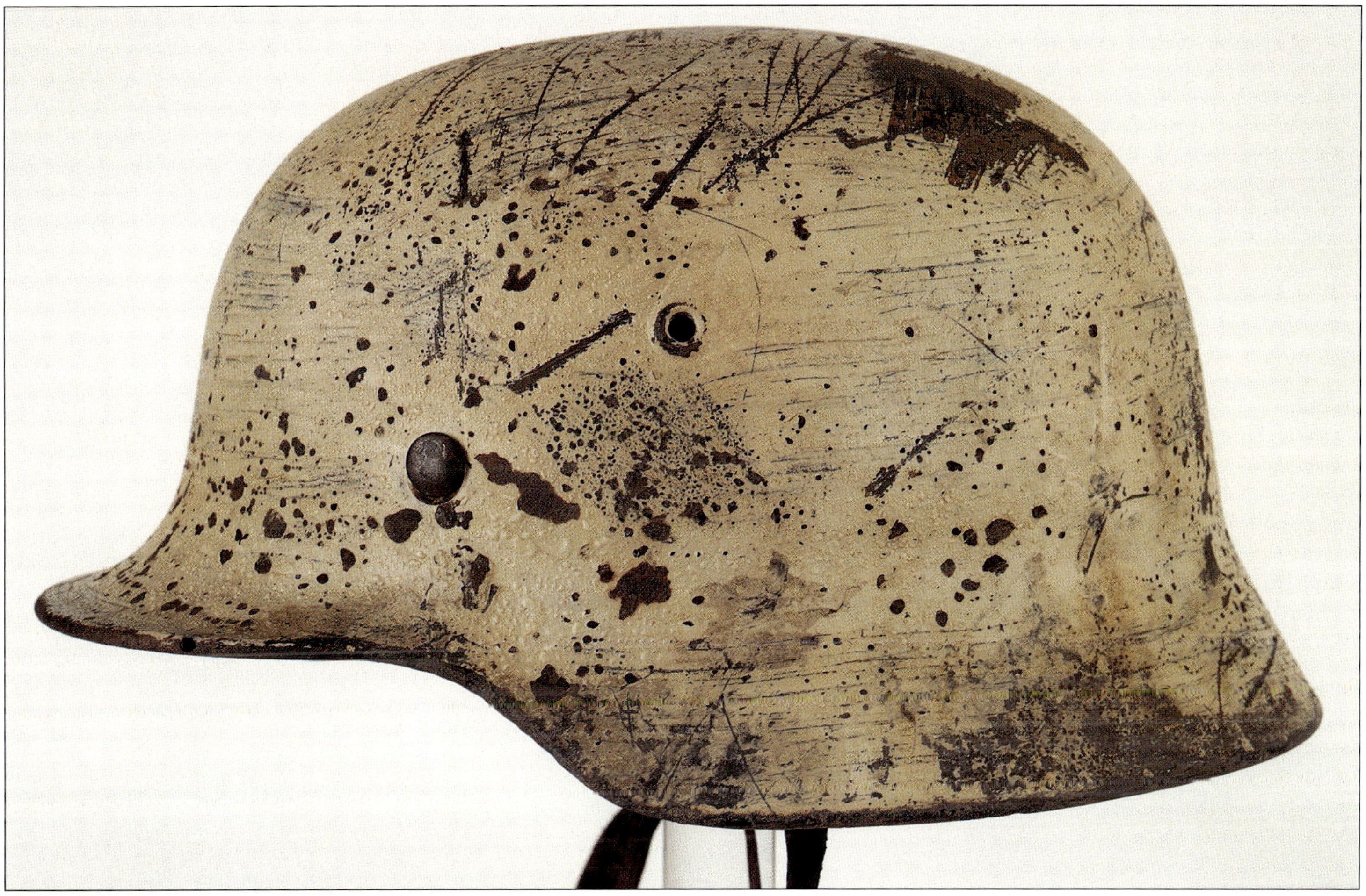

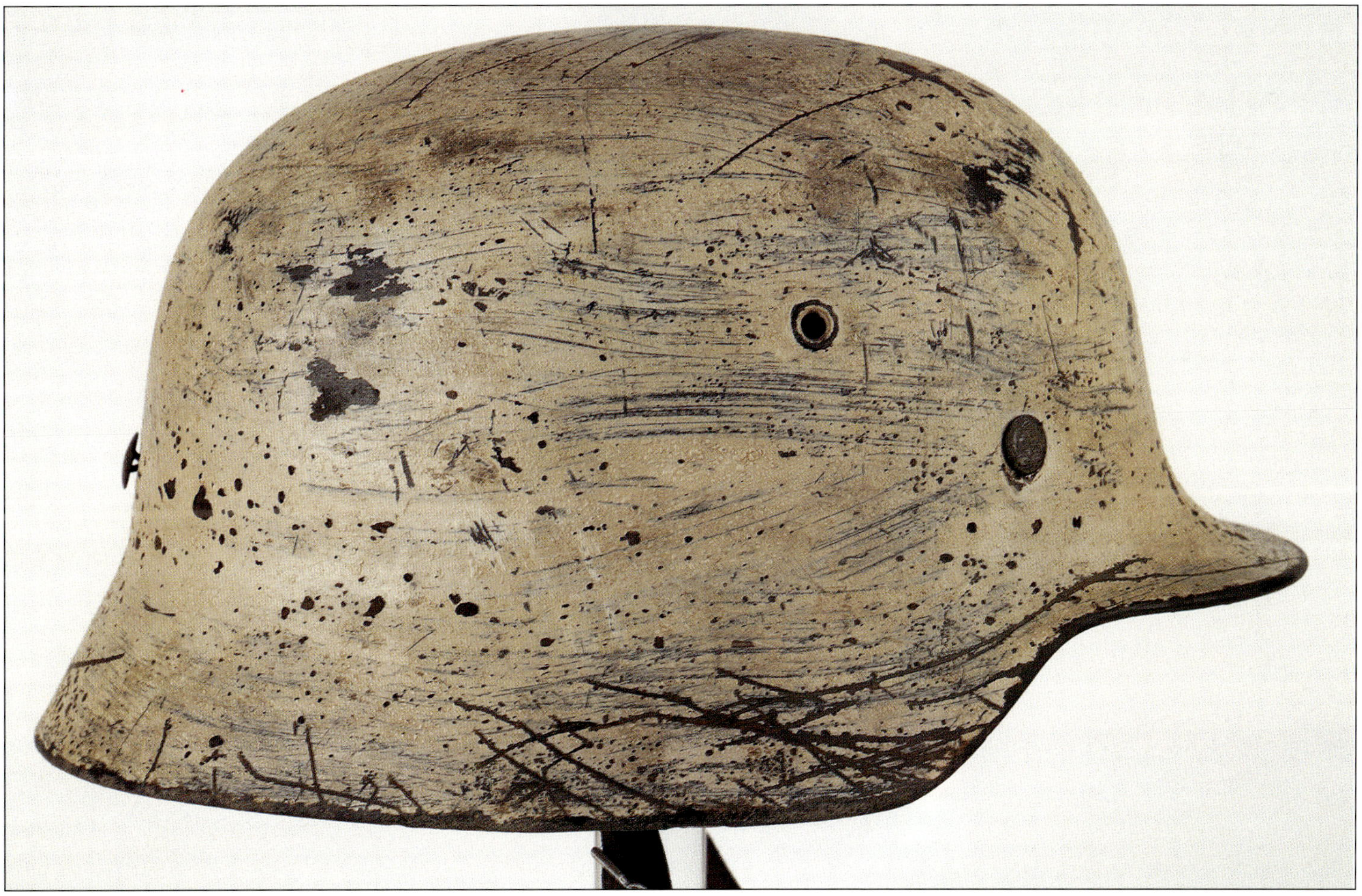

■ M40 helmet shell with M31 liner and chinstrap. This helmet was originally painted a matte field grey, then camouflaged with a sand colour. At one time it was covered in chicken wire and appears to have been painted in splotches of brown and green over the wire. No insignia is visible.

■ M40 helmet shell with M31 liner and chinstrap. This helmet was originally painted a matte field grey, then covered with a mixture of sand and sand-coloured paint, creating a paste-like texture. Then green has been brushed over the top. No insignia is visible.

■ M40 helmet shell with M31 liner and chinstrap. This helmet originally carried the Wehrmacht silver eagle transfer and was then entirely overpainted in a smooth sand camouflage. It was found in Crete in 1981.

M40 helmet shell with M31 liner and chinstrap. This helmet was originally painted a matte field grey and carried the single Wehrmacht silver eagle transfer. It was then overpainted with yellow, brown and green. It originally belonged to a man from 1 Flughafen Betriebes Kp. KG77 (squadron ground troops). It appears to have been re-issued after 1943 to a man in Stab 1 Einheit Fallschirm Artillerie Rgt. 5, 5 Fallschirmjäger Div. We know this because on the inside neck apron is a name and fieldpost number which in turn has been overpainted by another name and fieldpost number. Refer to the section on Owner-Named Helmets in the second volume.

■ M40 helmet shell with M31 liner and chinstrap. This helmet was originally painted a matte field grey and carried the single Wehrmacht silver eagle transfer, then later overpainted in a smooth sand colour. It was taken from a grave marker in Tunisia in 1943 by a British soldier.

An interior shot showing the unusual pigskin liner which the author has noted on several 'tropical' camouflaged helmets. A label is glued to the inside of the helmet neckguard stating where and when the helmet was found and a photograph of the helmet in its original position atop a grave marker.

■ The envelope and reverse of the photograph with an inscription which reads: 'Between Bardia & Gambut German - Helmet Souvenir' and is stamped as having passed by the Censor on 10th June 1943.

M40 helmet shell with M31 liner and chinstrap. This helmet was originally painted a matte field grey and carried the single Wehrmacht silver eagle transfer. It was then painted a base sand colour and sprayed green and brown over the top. This type of camouflage is often called 'Normandy style' but in fact this helmet was captured in Tunisia in 1943. (Collection of David Hucklebridge)

M40 helmet shell with M31 liner and chinstrap. This helmet was originally painted a matte field grey, then daubed with rag or sponge in red, brown, yellow and green. No insignia is visible.

M40 helmet shell with M31 liner and chinstrap. This helmet was originally painted a matte field grey, then overpainted in white. It carries the earlier pattern SS runic shield on the right side, and is named to R. Nitsch SS Uscha.

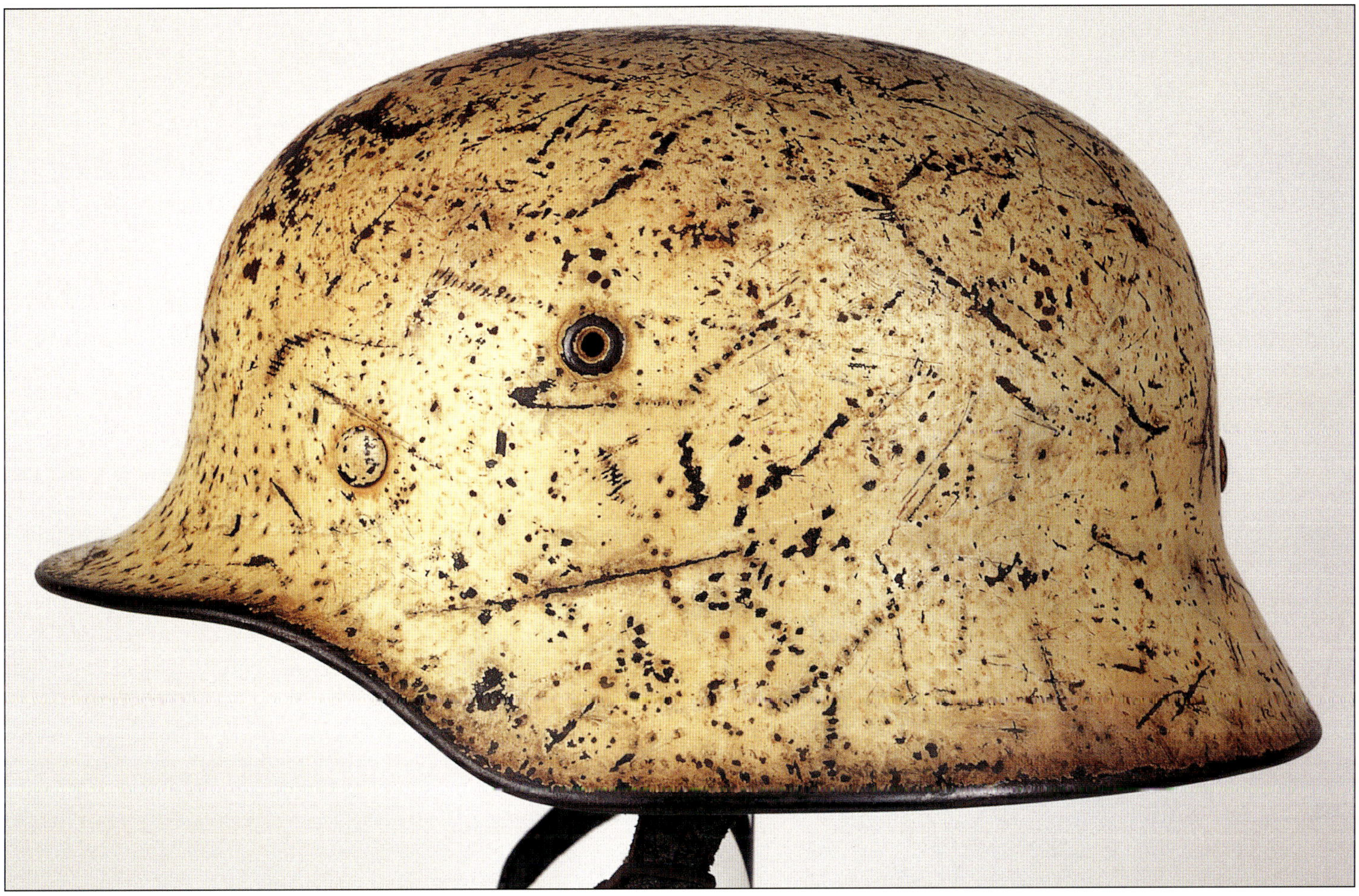

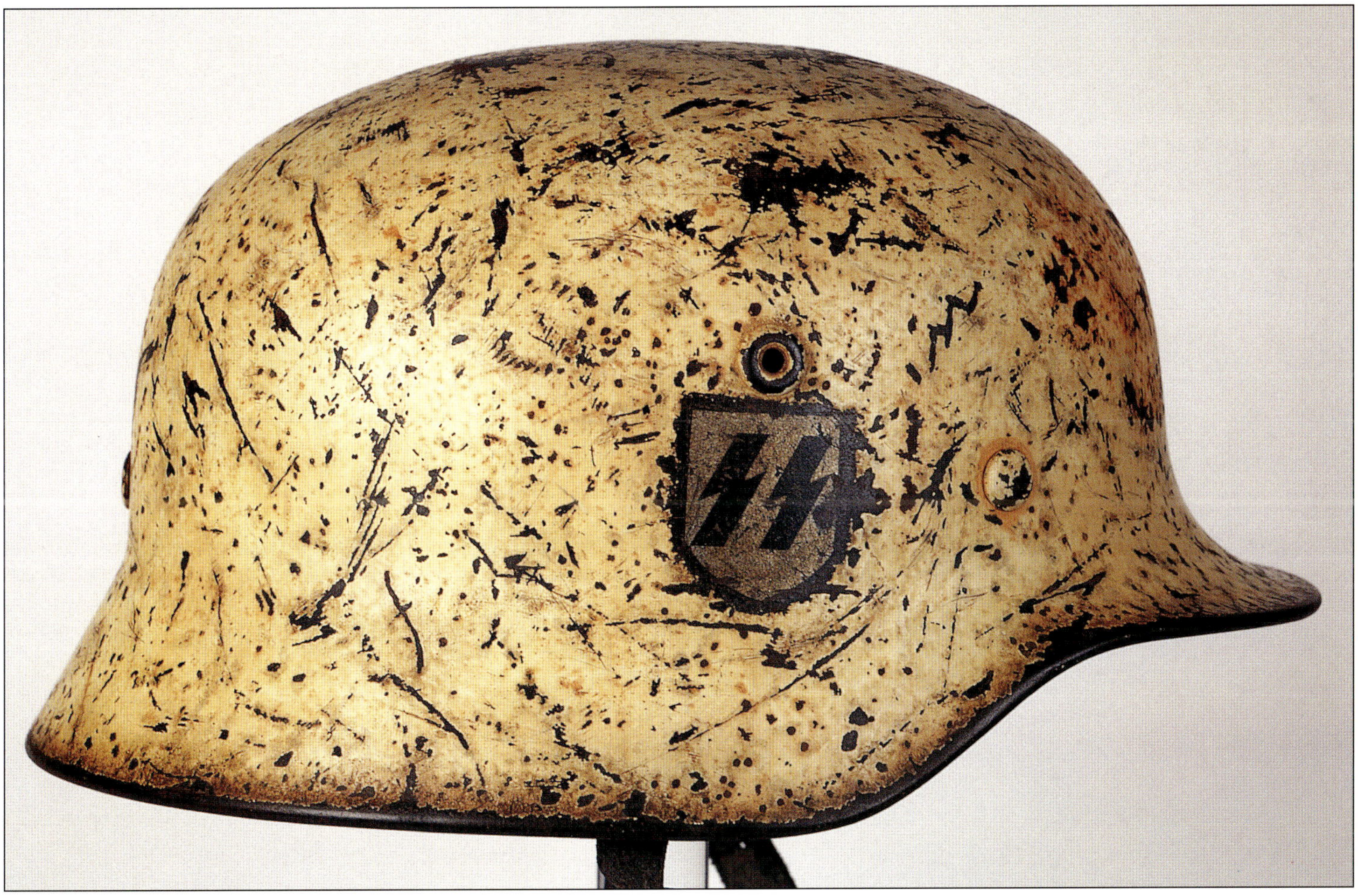

M40 helmet shell with M31 liner and chinstrap. This helmet was originally painted a matte field grey. It carries the earlier pattern SS runic shield on the right side and has been repainted in green with patches of brown and is named to SS Pionier Deutschmann.

M40 helmet shell with M31 liner and chinstrap. This helmet was originally painted a matte field grey and overpainted in white. It carries the earlier pattern SS runic shield on the right side and the Army-style Croatian shield transfer on the left.

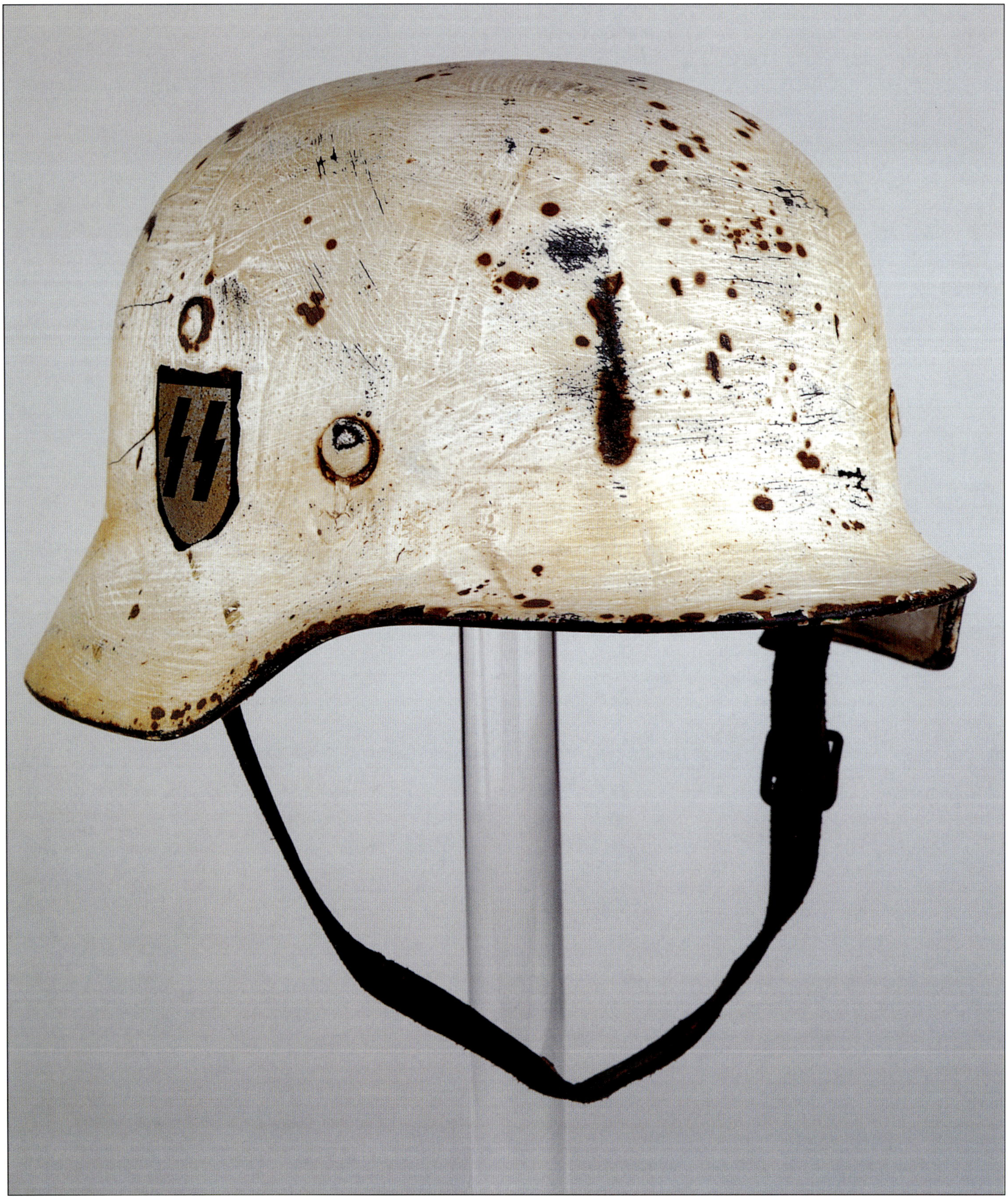

■ M40 helmet shell with M31 liner and chinstrap. This helmet was originally painted a matte field grey. It carries the later pattern SS runic shield on the right side and has been overpainted in whitewash, which could be scrubbed off, if required.

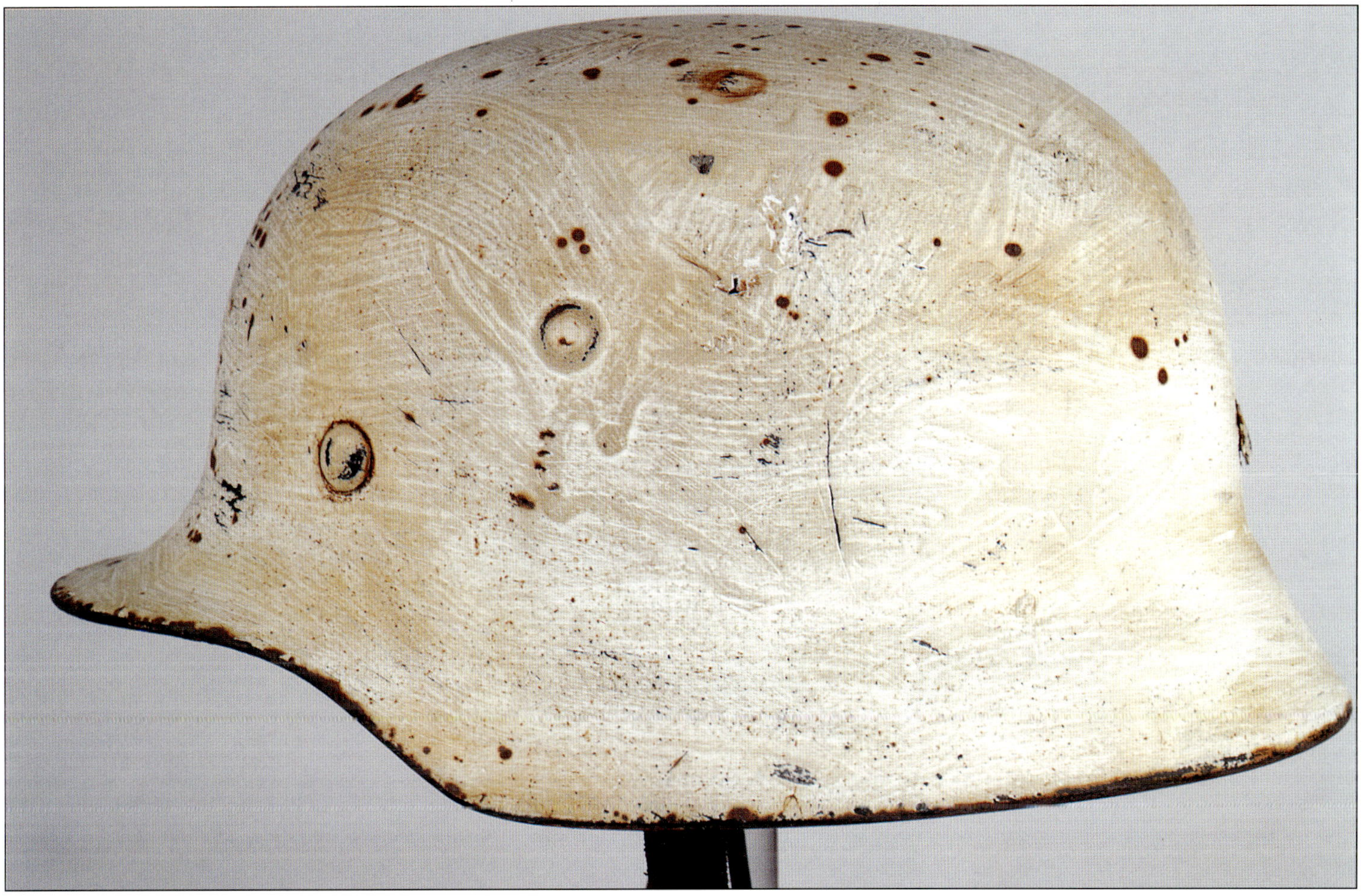

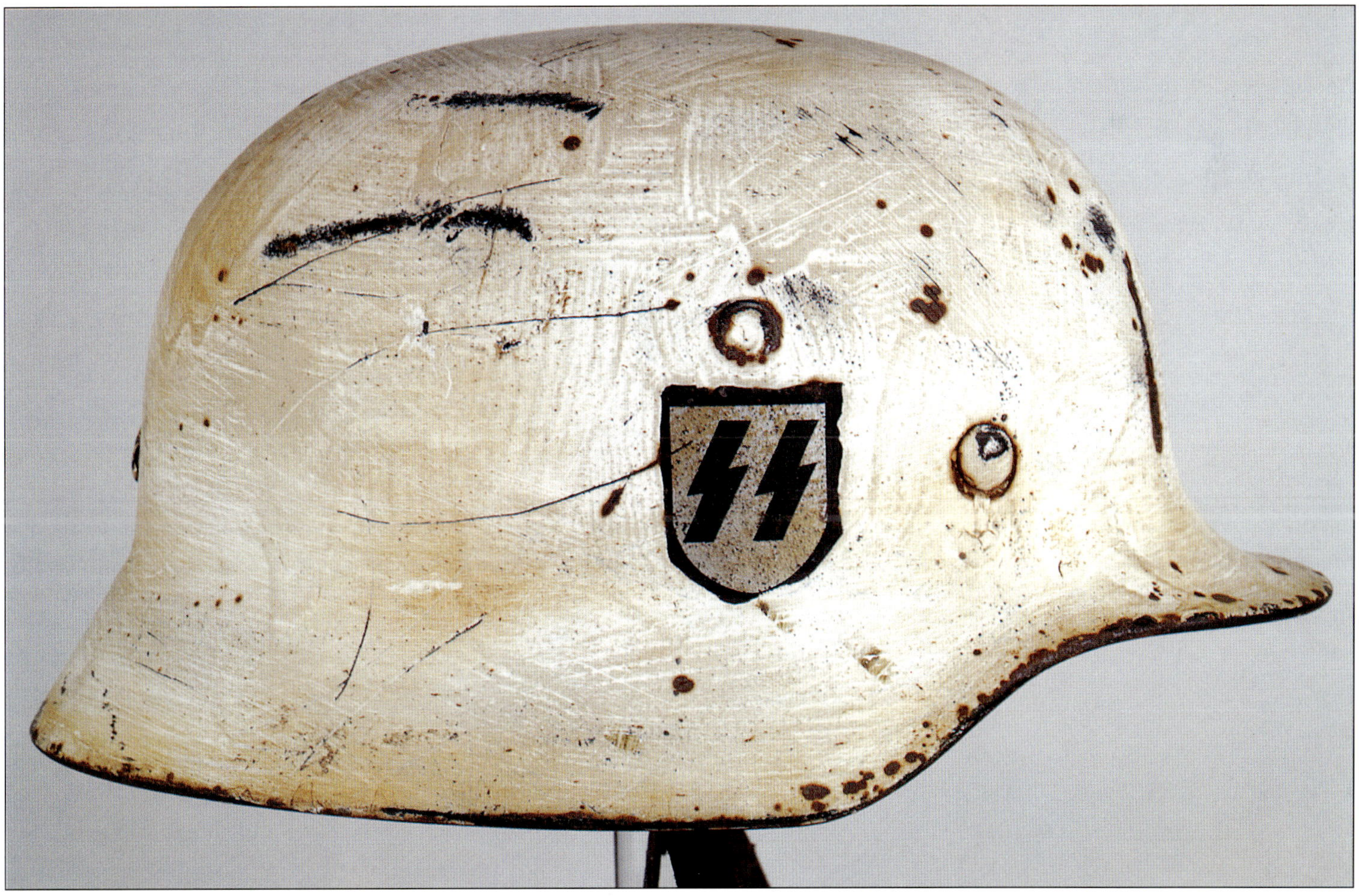

M40 helmet shell with M31 liner and chinstrap. This helmet was originally painted a matte Luftwaffe blue-grey. It carries traces of the later pattern Luftwaffe eagle on the left side and has been overpainted with a base sand colour and then sprayed brown and green, a typical example of the Normandy camouflage, where it was found. In the liner had been placed a German army newspaper from Norway.

M40 helmet shell with M31 liner and chinstrap. This helmet was originally painted a matte Luftwaffe blue-grey. It carries the later pattern Luftwaffe eagle on the left side and has been overpainted with a base sand colour and then sprayed brown and green. Note the placing of the Luftwaffe transfer on top of the new camouflage scheme over the original transfer, also late pattern, the outline of which can be seen quite clearly. This shows that frontline troops were sometimes able to acquire new transfers.

M40 helmet shell with M31 liner. This helmet was originally painted a matte Luftwaffe blue-grey. It carries the later pattern Luftwaffe eagle on the left side. It has been repainted a base sand colour around the insignia, then overpainted in brown and dull green.

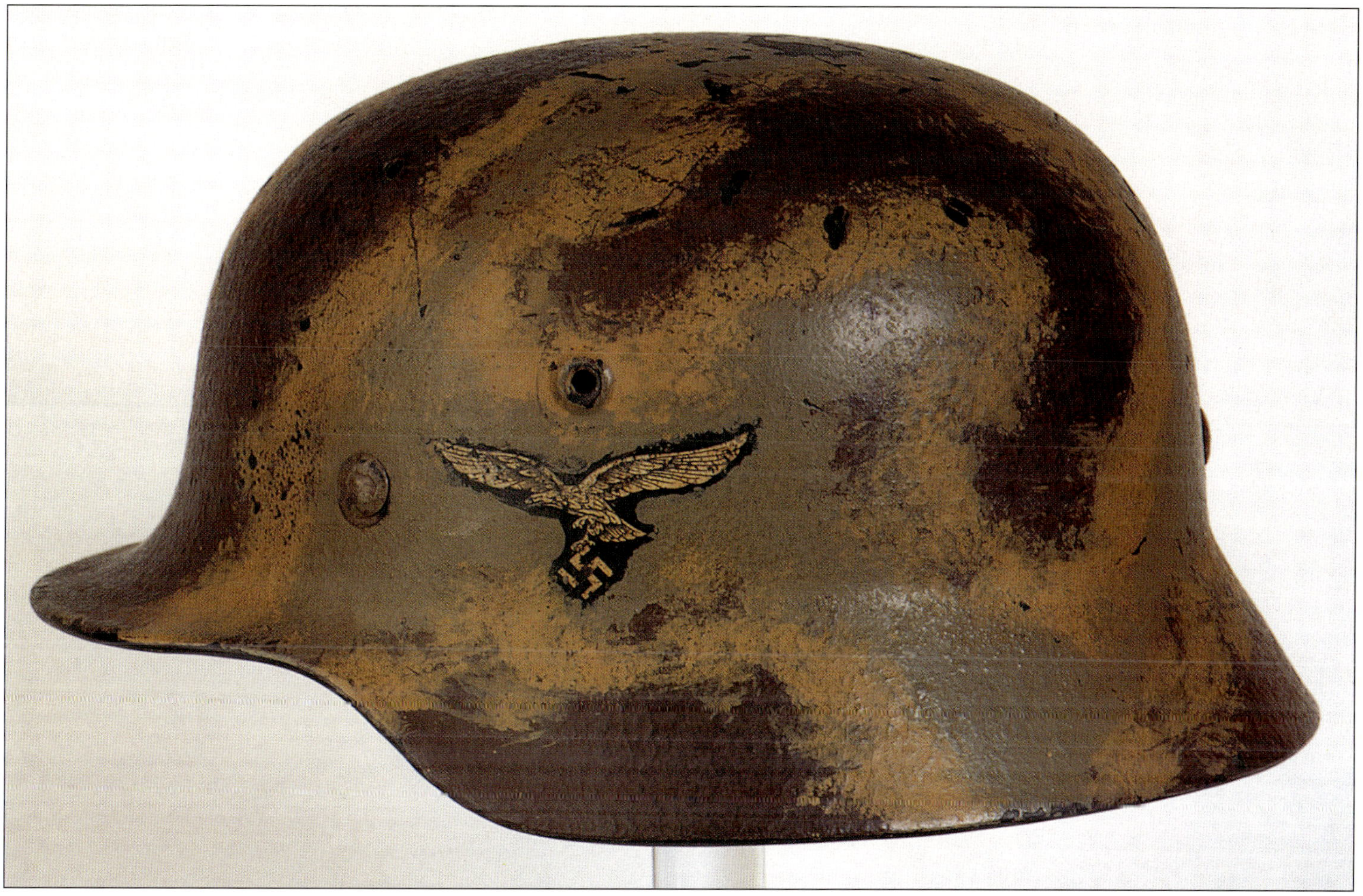

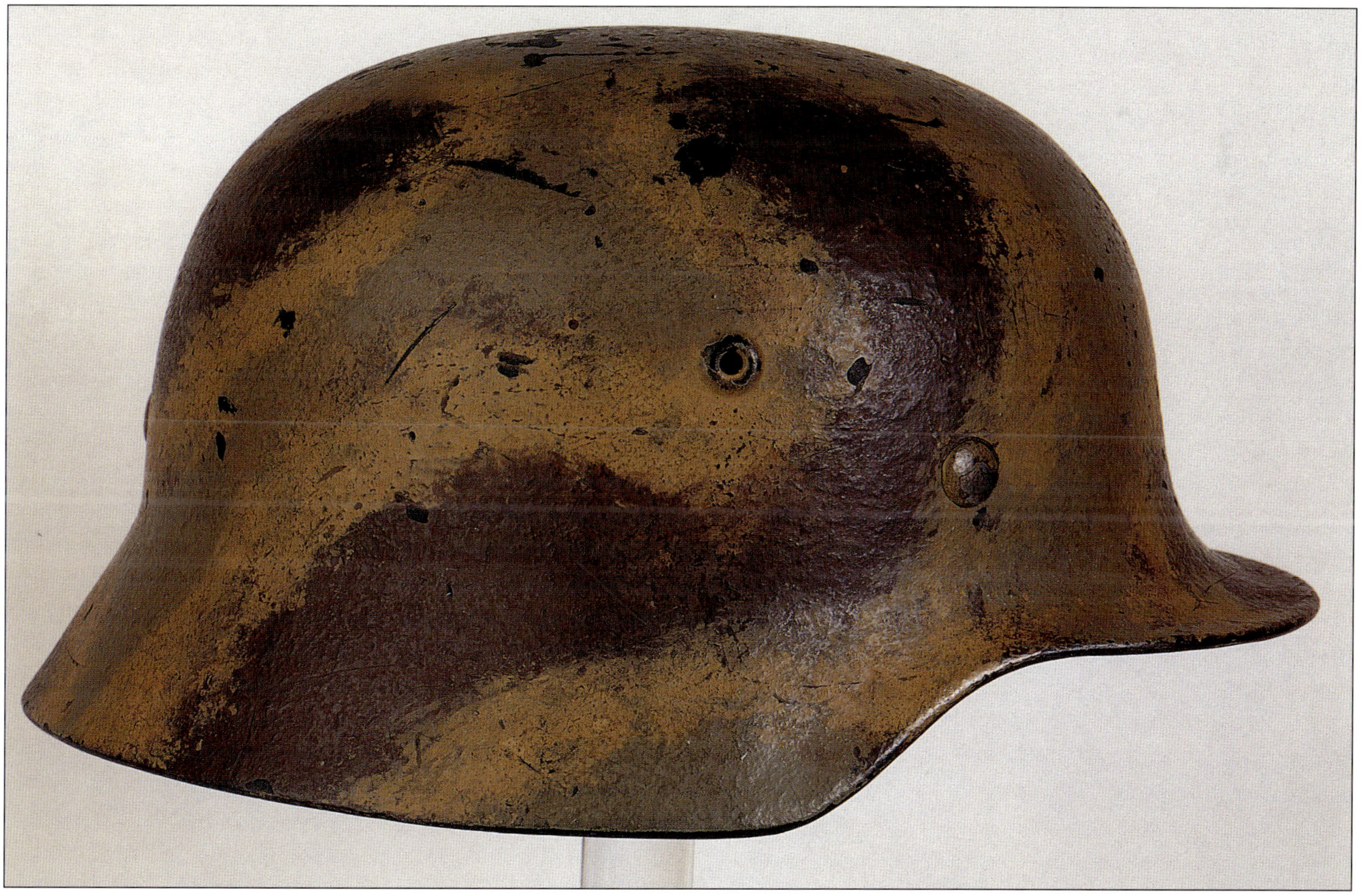

■ M40 helmet shell with M31 liner and chinstrap. This helmet was originally painted a matte Luftwaffe blue-grey. It carries the twin insignia of the Luftwaffe: the late pattern subdued eagle transfer on the left and national tri-colour shield transfer on the right. These transfers have been applied over the camouflage. (Collection of Marcus Cotton)

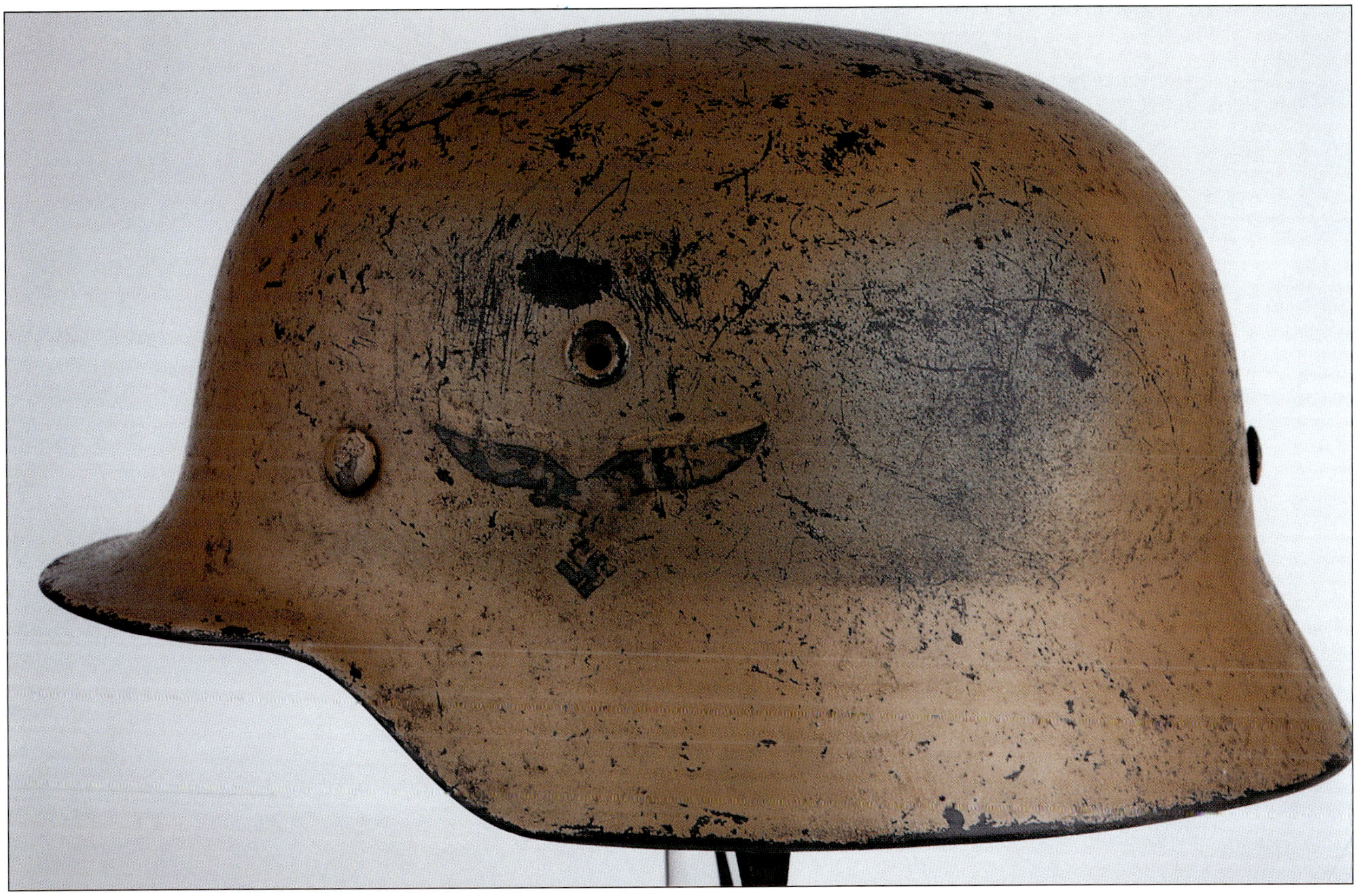

M40 helmet shell with M31 liner and chinstrap. This helmet was originally painted a matte Luftwaffe blue-grey. It carries the later pattern Luftwaffe eagle on the left side and has been crudely brushed over with sand-coloured paint. (Collection of Bill Shea - The Ruptured Duck)

■ M40 helmet shell with M31 liner and chinstrap. This helmet was originally painted a matte Luftwaffe blue-grey. It carries the later pattern Luftwaffe eagle on the left side and has been camouflaged with a base of green with black smudges over the top. Two things are of interest here: the printed paper skull stuck on the front and the numbers 2782 stenciled on the left rear. The significance of the skull is unknown; the serial number could relate to the owner. This helmet was found in an attic in Paris.

M40 helmet shell with M31 liner and chinstrap. This helmet was originally painted a matte Luftwaffe blue-grey. It carries the remnants of the later pattern Luftwaffe eagle on the left side and has been overpainted in a base sand colour, then crudely brushed over with white paint. According to the label inside, the owner of this helmet was a member of a searchlight crew.

M40 helmet shell with M31 liner. This helmet was originally painted a matte field grey. It carries the Wehrmacht eagle transfer on the left side and has been crudely covered in chicken wire to carry foliage.

M40 helmet shell with M31 liner and chinstrap. This helmet was originally painted a matte Luftwaffe blue-grey. It carries no insignia but has been covered in chicken wire.

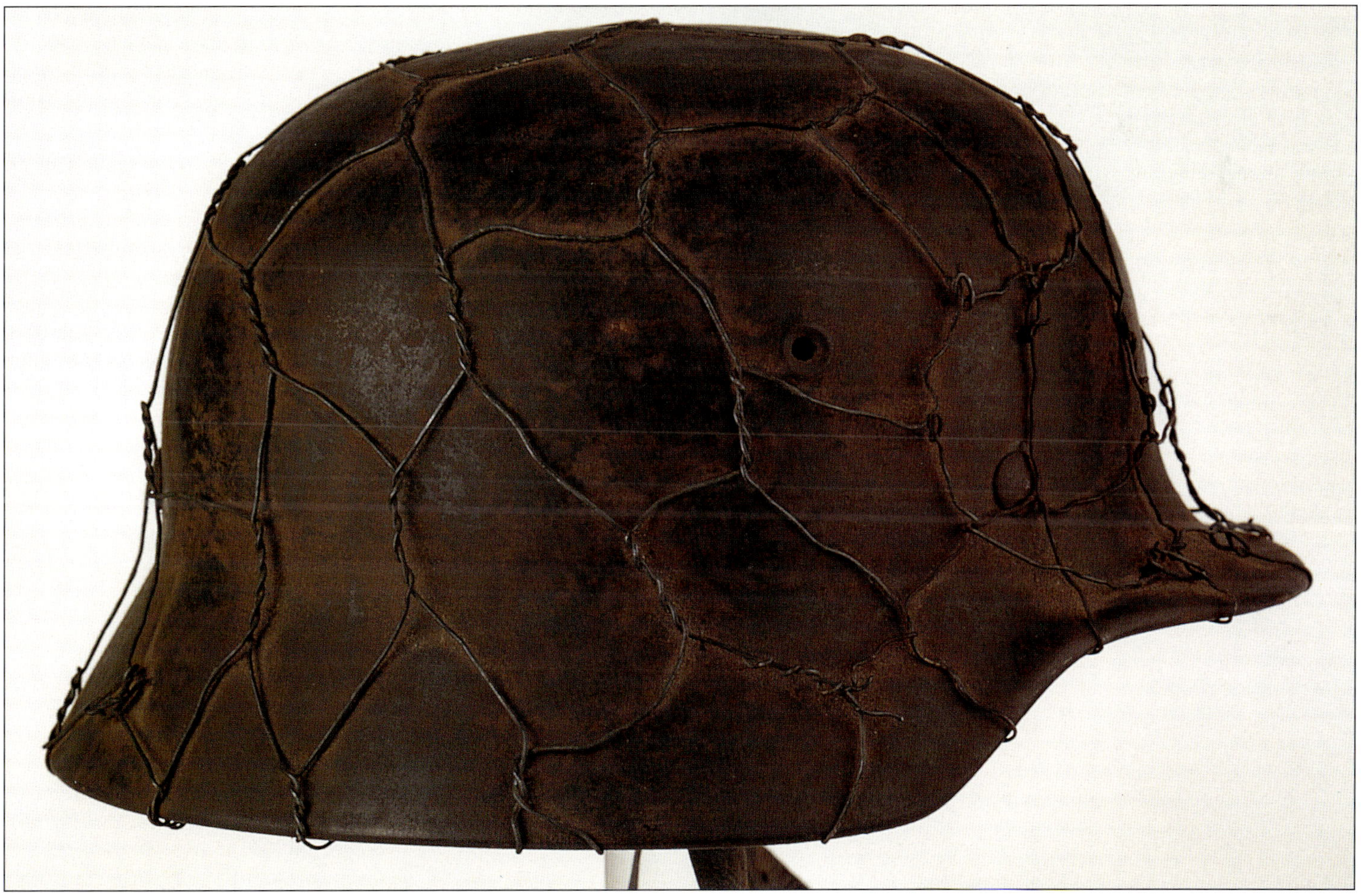

■ M40 helmet shell with M31 liner and chinstrap. This helmet was originally painted a matte field grey and has been overpainted with a base sand colour, then sprayed in green and brown patches. An elaborate wire and chicken wire cover has been placed over the top with interesting loops across the front. (Collection of Marcus Cotton)

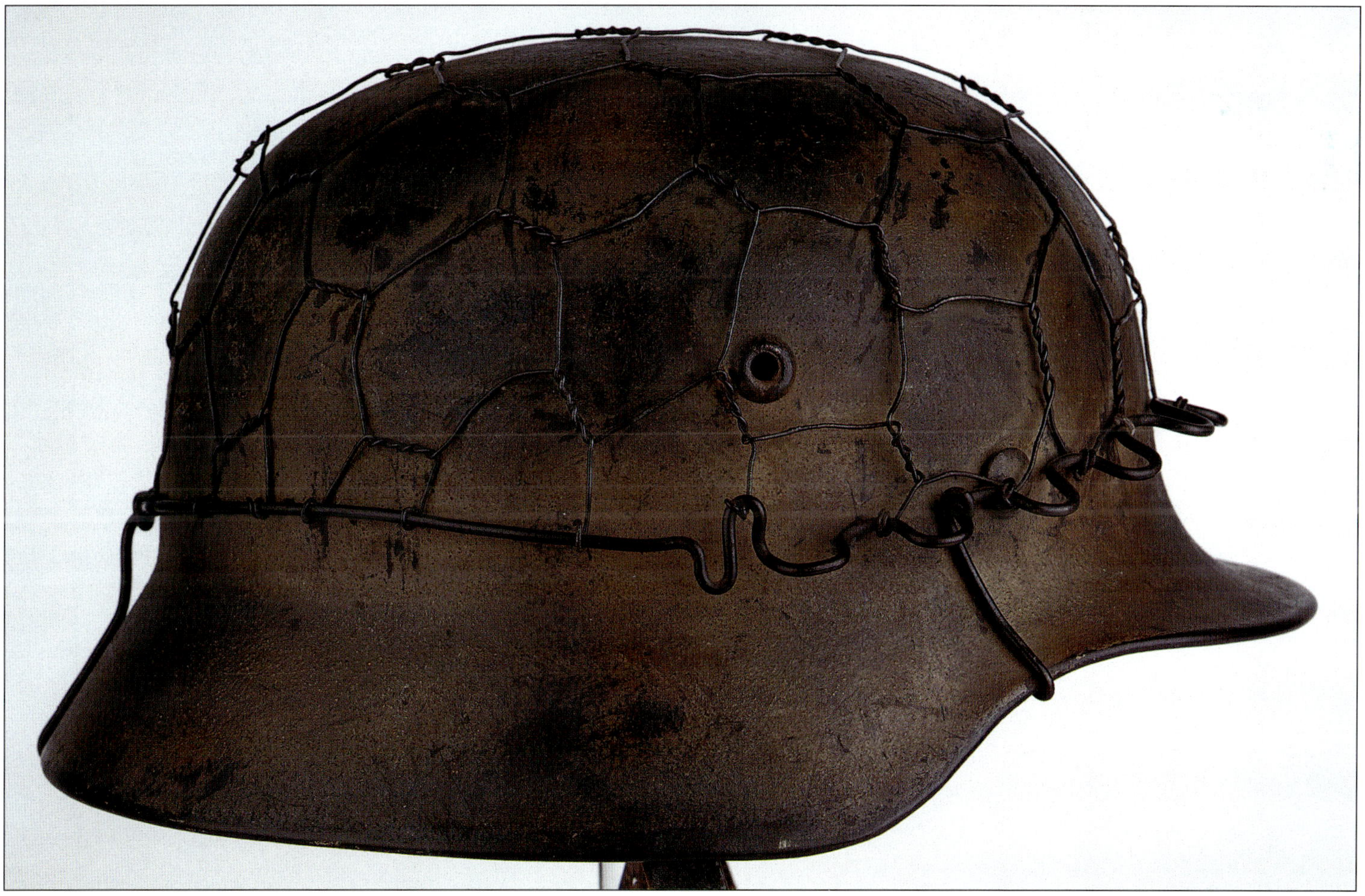

M40 helmet shell with M31 liner and chinstrap. This helmet is painted a matte Luftwaffe blue-grey. It carries the later pattern Luftwaffe eagle on the left side then covered with wire. (Collection of Bill Shea - The Ruptured Duck)

M40 helmet shell with M31 liner and chinstrap. This helmet is painted a matte field grey. It carries the twin insignia of the Feldpolizei and has been covered with chicken wire.

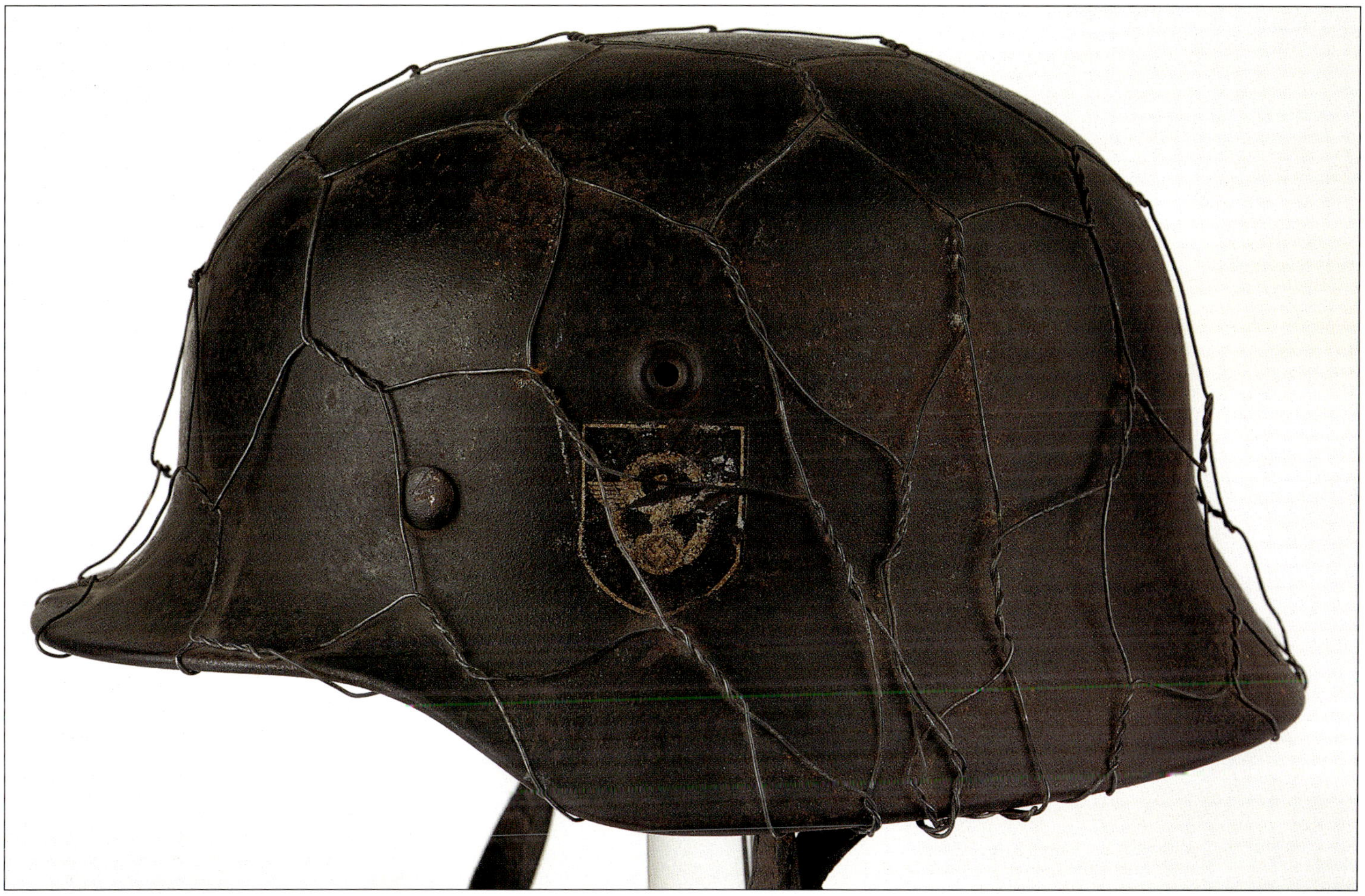

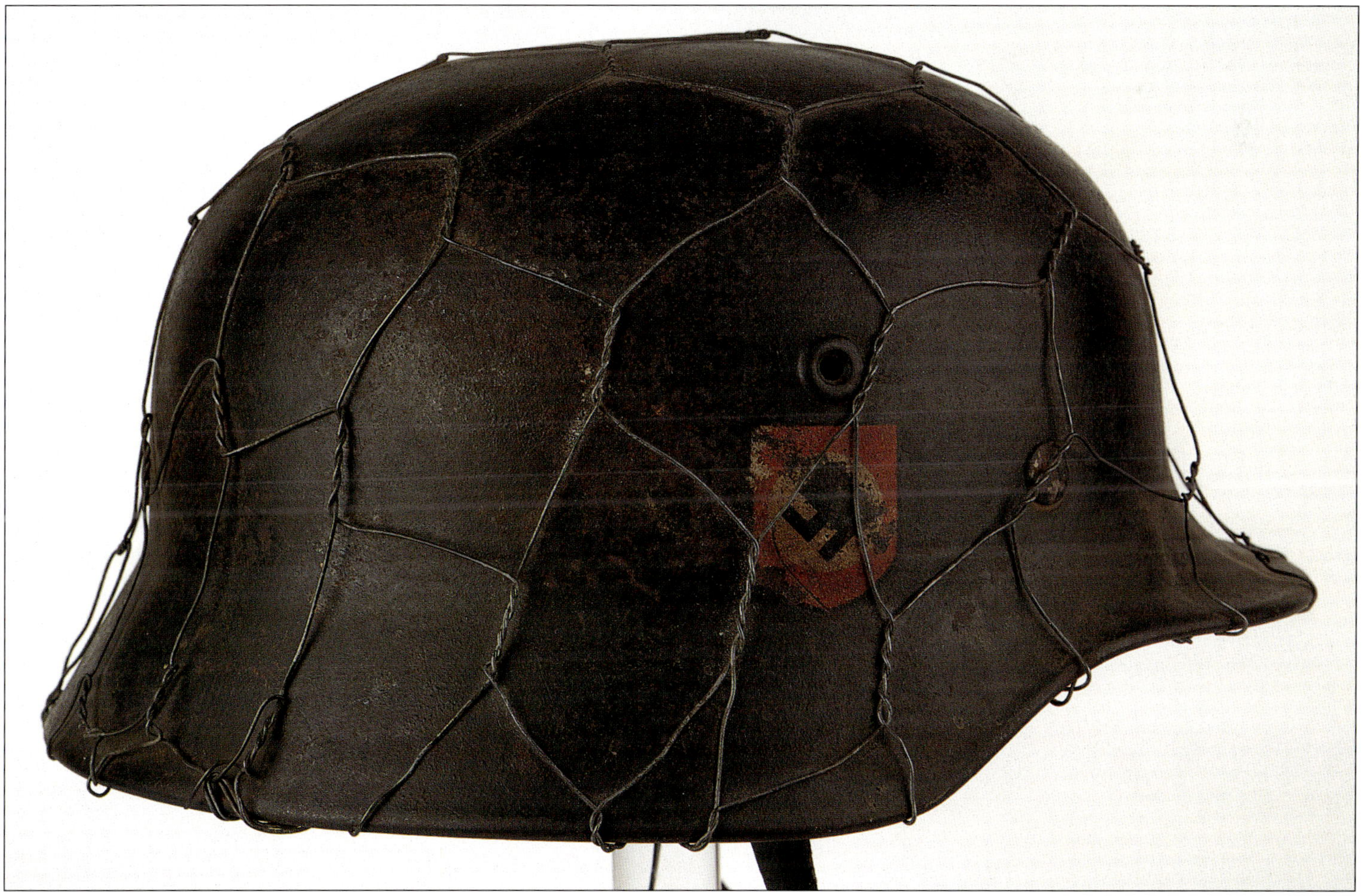

M40 helmet shell with M31 liner and chinstrap. This helmet is painted a satin apple-green (now darkened with age). It carries the later pattern SS runic shield transfer on the left side and the swastika shield transfer on the right. It is covered with wire for foliage. Usually when SS insignia are found reversed, it is because the helmet was originally used by the Feldpolizei and its eagle transfer has been covered by the SS runic shield. This happened when the 4th SS Polizei Division exchanged their police-style uniforms for SS ones. However, this is not so in this case. I would suggest this is either a mistake during initial application or an affectation by its owner.

M40 helmet shell with M31 liner and chinstrap. This helmet is painted a matte Luftwaffe blue-grey. It carries the later pattern Luftwaffe eagle on the left side and has a cut down version regulation camouflage string netting attached. (Collection of Malcolm Fisher - Regimentals)

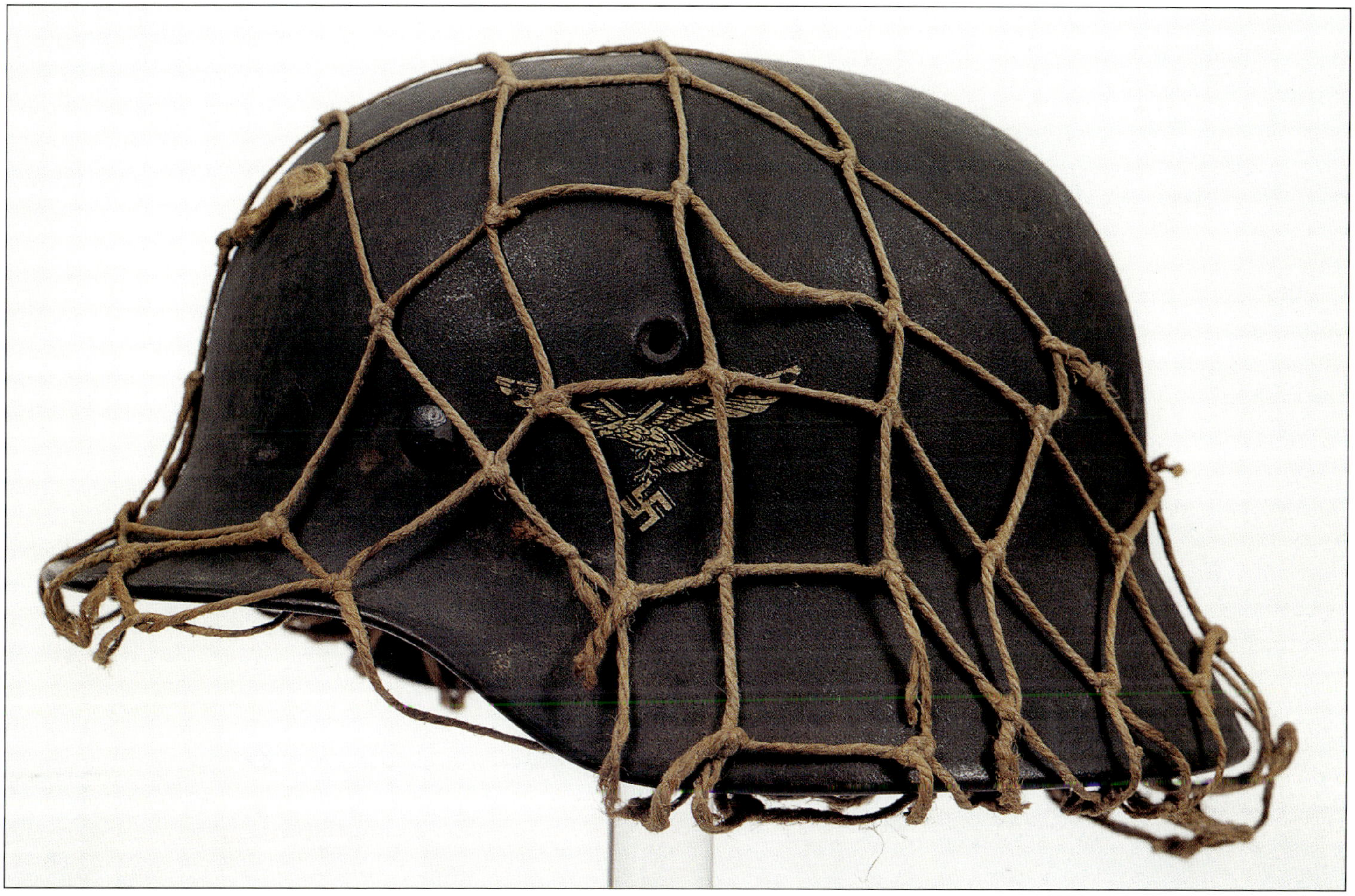

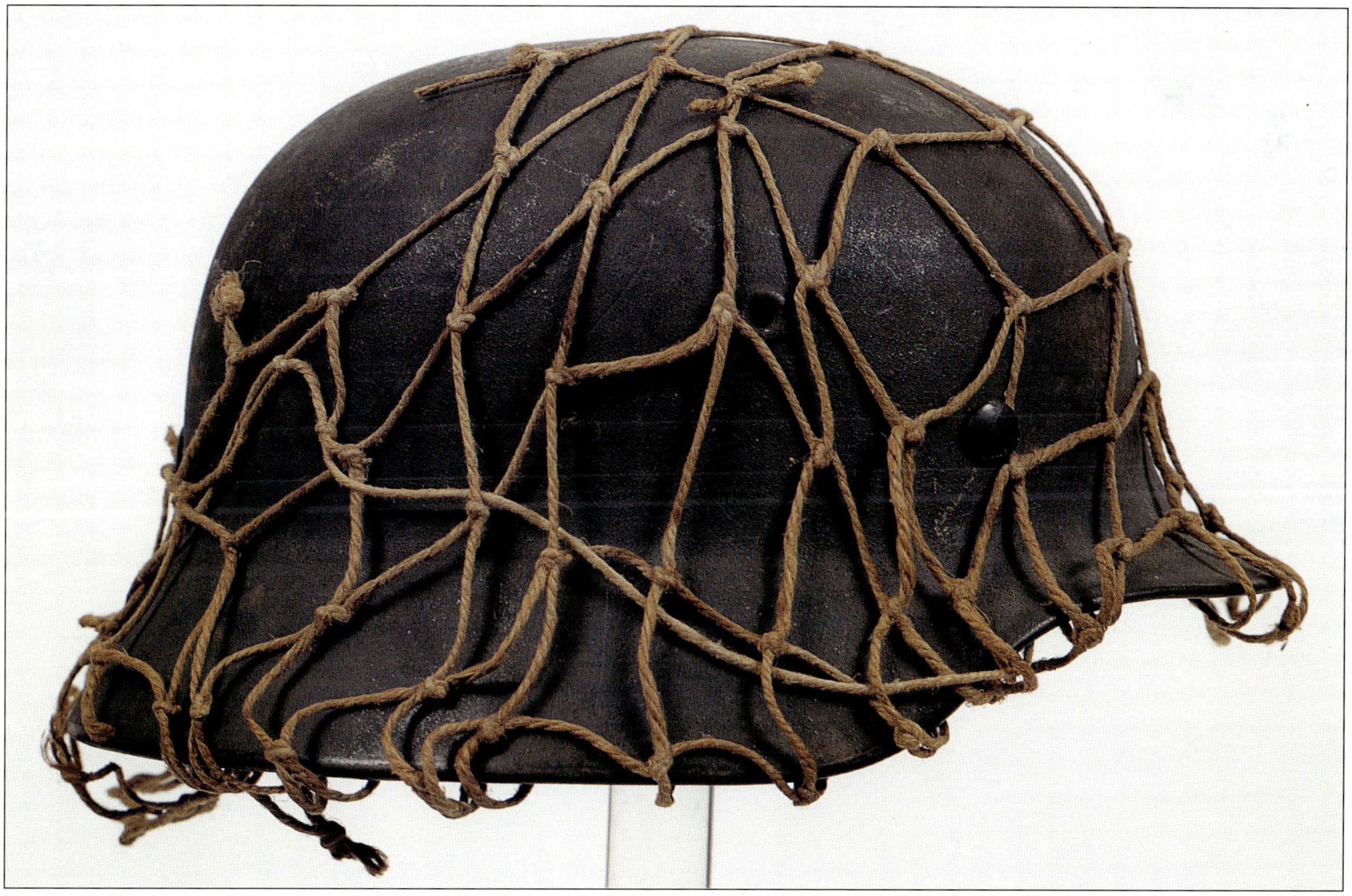

M40 helmet shell with M31 liner and chinstrap. This helmet was originally painted a matte Luftwaffe blue-grey and has been crudely overpainted in sand coloured camouflage. It carries the remains of the later pattern Luftwaffe eagle on the left side and has the full regulation camouflage string netting attached. This helmet was found recently in loft space over a garage in the USA.

Front and back views.

Interior view.

MODEL 1942

■ M42 helmet shell with M31 liner and chinstrap. This helmet is painted a smooth field grey, made shiny by wear. It carries the Wehrmacht silver eagle transfer on the left side.

■ M42 helmet shell with M31 liner. This helmet is painted a matte field grey. It carries the Wehrmacht silver eagle transfer on the left side.

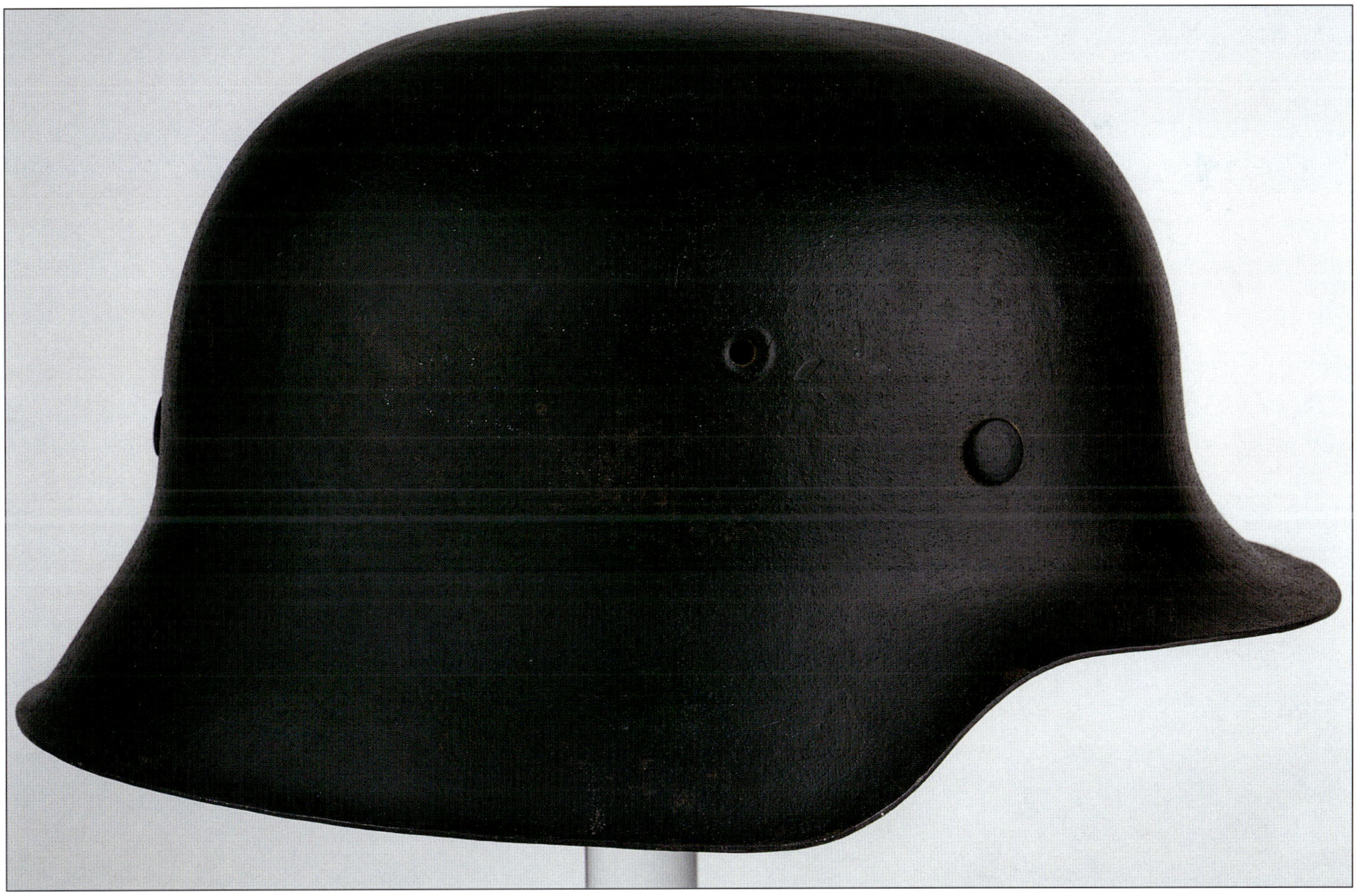

M42 helmet shell with M31 liner and chinstrap. This helmet is painted a matte naval grey, made shiny by wear. It carries the Kriegsmarine gold eagle transfer on the left side and the 9. U-Flotille (u-boat) insignia painted in red on the right. This is an interesting helmet. The grey is the only paint on the shell but strangely it seems to have been painted carefully around the transfer, indicating the possibility that the transfer was placed directly onto bare metal. Also the helmet shell itself has been crudely manufactured. As can be seen from the side views particularly, it lacks the usual curve in the drop from the visor to the neck guard. These unusual features suggest this is a very late-war helmet, although another strange anomaly is it is stamped 'ET' inside the left side of the neck guard, an early- to mid-war manufacturer's mark, and not 'ckl' at the back of the neck guard as was used towards the end of the war.

■ M42 helmet shell with M31 liner. This helmet is painted a matte Luftwaffe blue-grey. It carries the late pattern Luftwaffe eagle transfer on the left side.

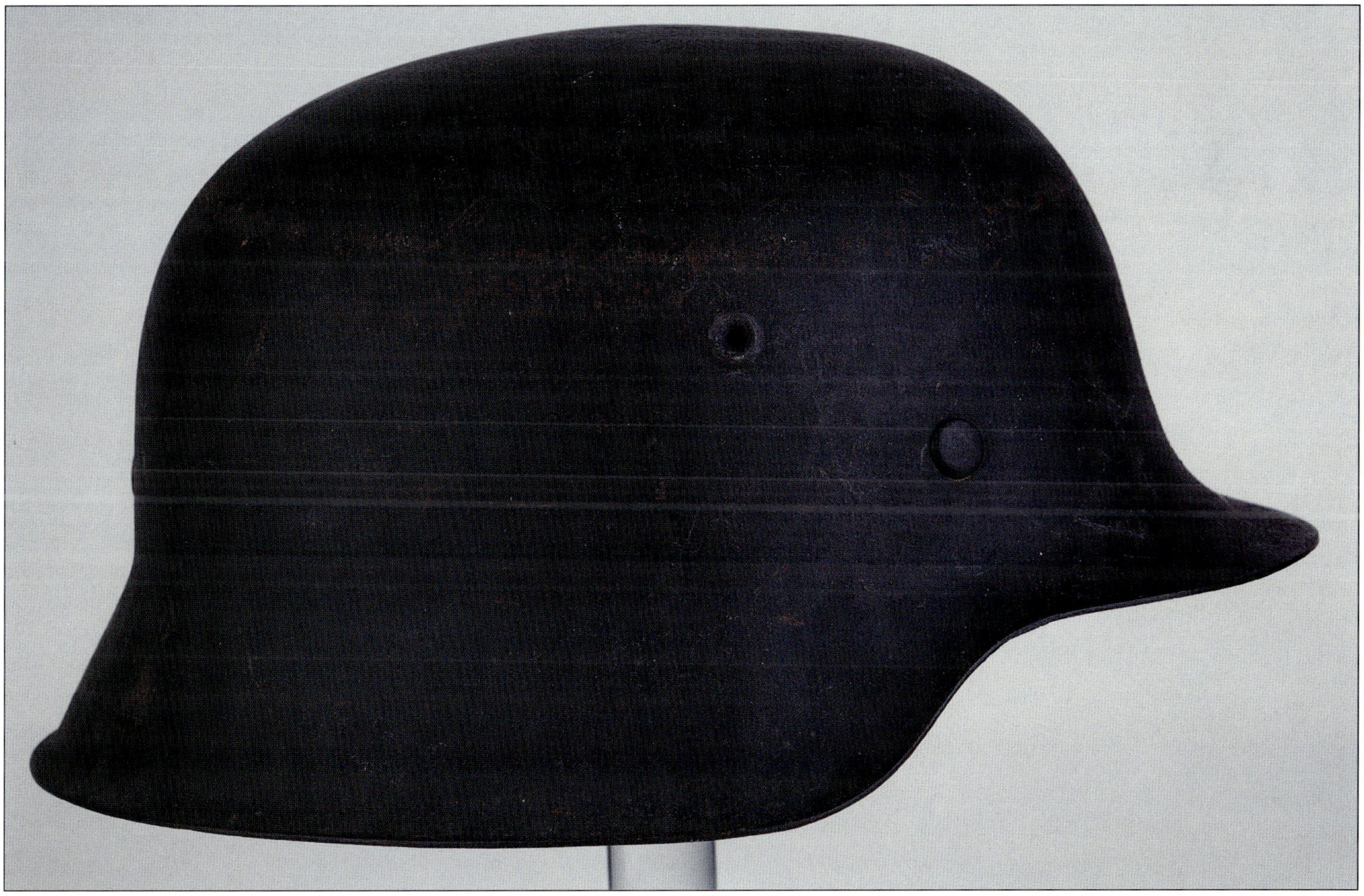

MODEL 1942

■ M42 helmet shell with M31 liner. This helmet is painted a matte Luftwaffe blue-grey. It carries the late pattern Luftwaffe eagle transfer on the left side. Shown clearly are the late-war aluminium split cotter pins. (Collection of Malcolm Fisher - Regimentals)

M42 helmet shell with M31 liner and chinstrap. This helmet is painted a matte field grey, made smooth with wear. It carries the earlier SS runic shield transfer on the right side. The chinstrap is buckled on the right side, presumably to allow the left-handed owner to sight and fire his rifle. Of note is the use of the so-called early-style transfer on a late-war helmet. (Collection of Malcolm Fisher - Regimentals)

■ M42 helmet shell with M31 liner. This helmet is painted a matte field grey, made smooth with wear. It carries the late-style SS runic shield transfer on the right side.

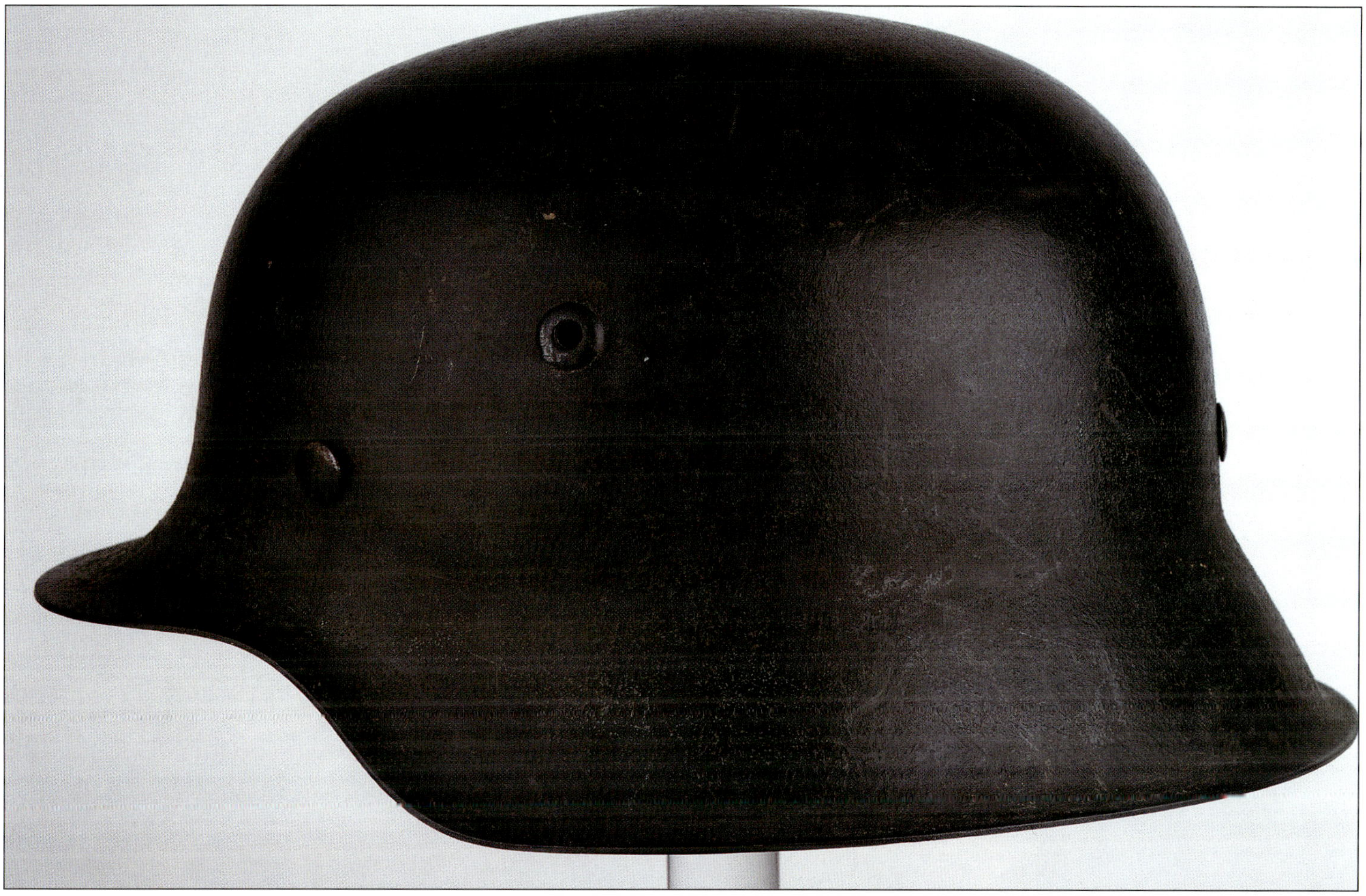

M42 helmet shell with M31 liner and chinstrap. This helmet is painted a matte field grey. It carries the single insignia of the 'Freikorps Sauerland' as a transfer on the left side. This was a Volksturm-type (home guard) unit raised at the end of 1944 in Sauerland with many of its members being factory workers and teachers. This helmet is an un-issued condition. A version of this helmet with the SS runic shield on the right side is in the West Point Museum in the USA.

Sauerland

■ M42 helmet shell with M31 liner and chinstrap. This helmet is painted a matte field grey and was produced without insignia.

■ M42 helmet shell with M31 liner and chinstrap. This helmet is painted a matte field grey, made smooth with wear. It carries the, possibly unique, insignia of the Freies Indien as a transfer on the left side and the Wehrmacht eagle transfer on the right. Normally, a plain tri-colour shield transfer of orange, white and green was worn on the left. This unit of volunteers for the German Army was drawn largely from Indian POWs captured in North Africa. They were transferred to SS control towards the end of the war. This helmet was found in Holland where some of them were based. It is interesting to note that the Army transfer has been placed on the right side.

FREIES INDIEN

■ M42 helmet shell with M31 liner and chinstrap. This helmet is painted a matte field grey. It carries the SS-style Croatian shield transfer on the left side and, again, the unusual placing of the Wehrmacht eagle transfer on the right.

M42 helmet shell with M31 liner and chinstrap. This helmet is painted a matte grey colour, made smooth with wear. It carries the painted shield of the Spanish Blue Division on the right side and the Wehrmacht eagle transfer on the left.

■ M42 helmet shell with M31 liner and chinstrap. This helmet is painted a matte Luftwaffe blue-grey and, unusually, it carries the painted insignia of 7./St.G 1. This unit, originally designated as 1./Tr.G 186, was formed as part of the strike component of the aircraft carrier Graf Zeppelin. As is well known, the carrier was never finished and this unit became the 7th Staffel in July 1940, flying Ju 87s. This helmet was found by an American soldier on a German air base in 1945.

M42 helmet shell with M31 liner. This helmet is painted a matte field grey, made smooth with wear. It carries no transfers but has a Polish Home Army cloth band around the shell, pinned at the back. This was worn on captured helmets during the Warsaw uprising in the summer of 1944.

M42 helmet shell with home-made liner and chinstrap . This helmet is painted a matte black, made smooth with wear. It carries the hand-painted Polish Home Army flag on each side. This came from a veteran of the 1944 Warsaw uprising, now living in the UK. Most probably, this was either a helmet shell found in a burnt-out building or, worse, a helmet with a liner in a bloodied condition, as the liner has been crudely replaced with a blanket attached through the original cotter pin holes with hammered nails. The chinstrap is made from a narrow belt. (Collection of Marcus Cotton)

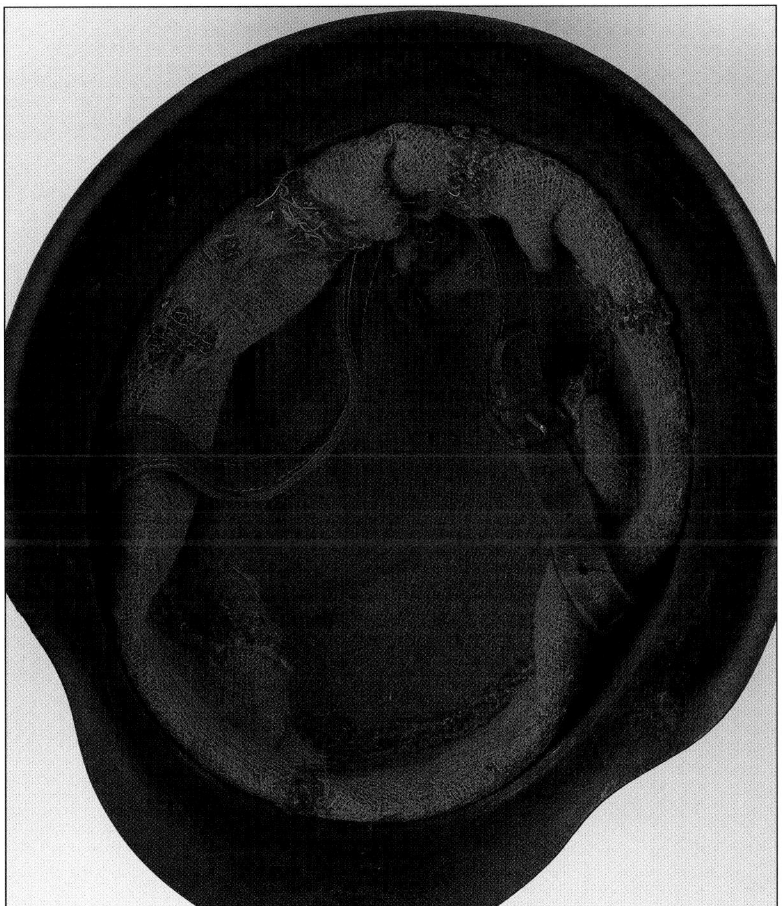

■ M42 helmet shell with M31 liner and chainstrap. This helmet is painted a matte field grey. It carries the insignia of the Ukrainian volunteers to the German Army, either the Ukrainian Liberation Army, the Ukrainian National Army or the 14th SS Ukrainian Division. This helmet was found in Prague, scene of the last battles for some of these units.

■ M42 helmet shell with M31 liner and chinstrap. This helmet has been repainted crudely in a subdued green. It carries the insignia of the Croatian 'Einsatzstaffel der Deutschen Mannschaft' (volunteer militia for the German Army) on the right side as a transfer. Examples do exist of this insignia on both sides or on one side with the Croatian shield on the other.

M42 helmet shell with M31 liner and chinstrap. This helmet was originally painted a matte field grey and has been overpainted with whitewash. It carries the Wehrmacht eagle transfer on the left side.

■ M42 helmet shell with M31 liner and chinstrap. This helmet was originally painted a matte field grey and has been completely overpainted with whitewash. No insignia is visible. This helmet was used on the Eastern Front and was recently brought back from Russia.

■ M42 helmet shell with M31 liner and chinstrap. This helmet was originally painted a matte field grey and has been overpainted with a base sand colour, with splashes of green and brown over the top. Traces of the Wehrmacht eagle transfer show through. This helmet was recently bought from a farmer in the Falaise area of Normandy who claimed to have picked it up in 1944.

MODEL 1942

M42 helmet shell with M31 liner and chinstrap. This helmet was originally painted a matte field grey and has been overpainted with a base colour of dark green. Leaf shapes in brown and yellow have been painted over the top. No insignia is visible. (Collection of Marcus Cotton)

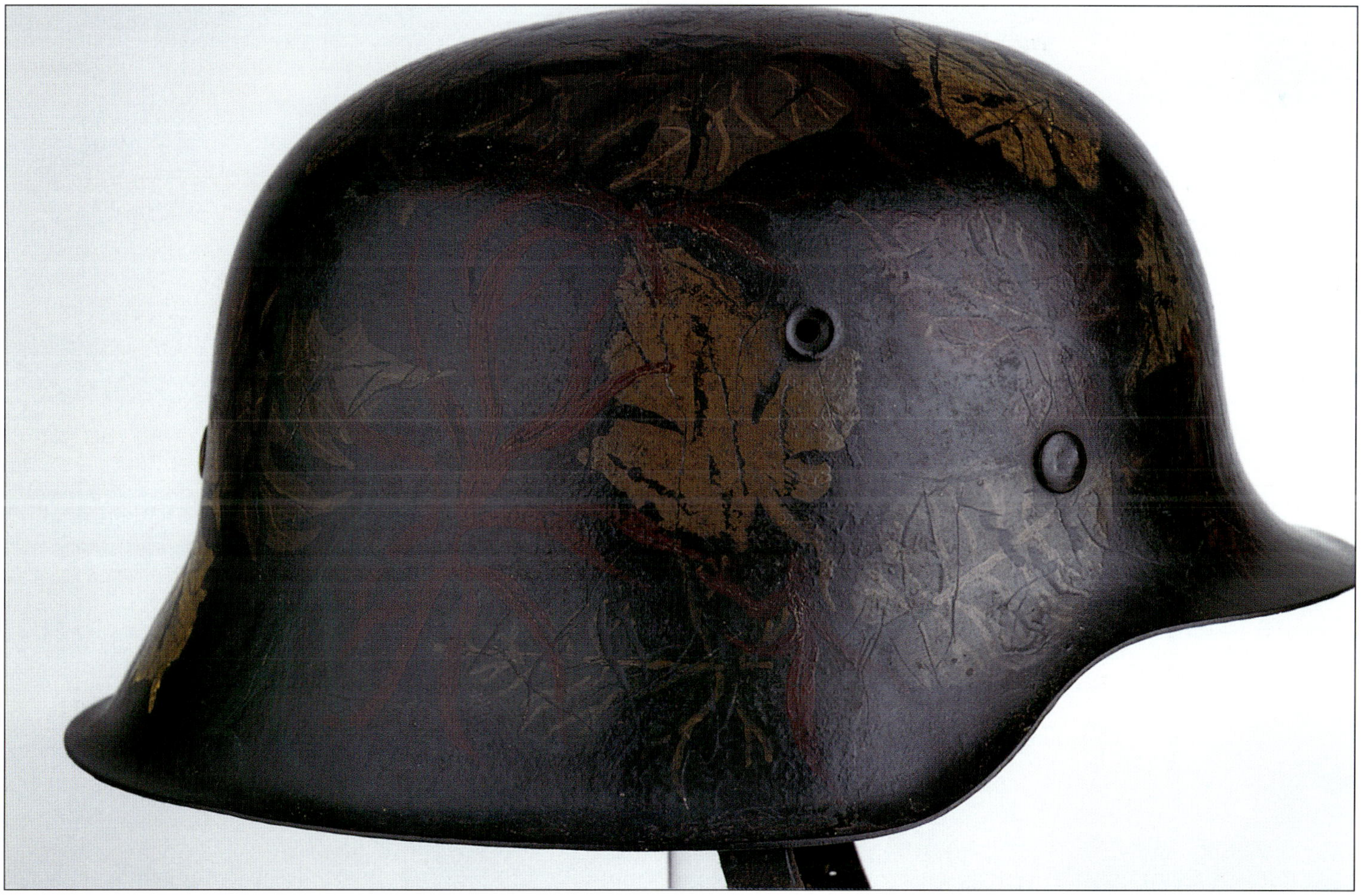

M42 helmet shell with M31 liner and chinstrap. This helmet was originally painted a matte field grey and has been overpainted with a sand base colour, then sprayed with patches of green. It carries the Wehrmacht eagle transfer on the left side and traces of the Red Cross eagle transfer on the right. (Collection of Malcolm Fisher - Regimentals)

■ M42 helmet shell with M31 liner and chinstrap. This helmet was originally painted a smooth field grey and has been overpainted with a sand base colour, then sprayed with patches of green and maroon/brown. No insignia is visible.

■ M42 helmet shell with M31 liner and chinstrap. This helmet was originally painted a matte field grey and has been overpainted with a sand base colour, then sprayed with patches of green and brown. No insignia is visible.

M42 helmet shell with M31 liner and chinstrap. This helmet was originally painted a matte Luftwaffe blue-grey and has been overpainted with a sand base colour, then sprayed with waves of green and brown. The outline of the late pattern Luftwaffe eagle transfer is visible under the paint. (Collection of Marcus Cotton)

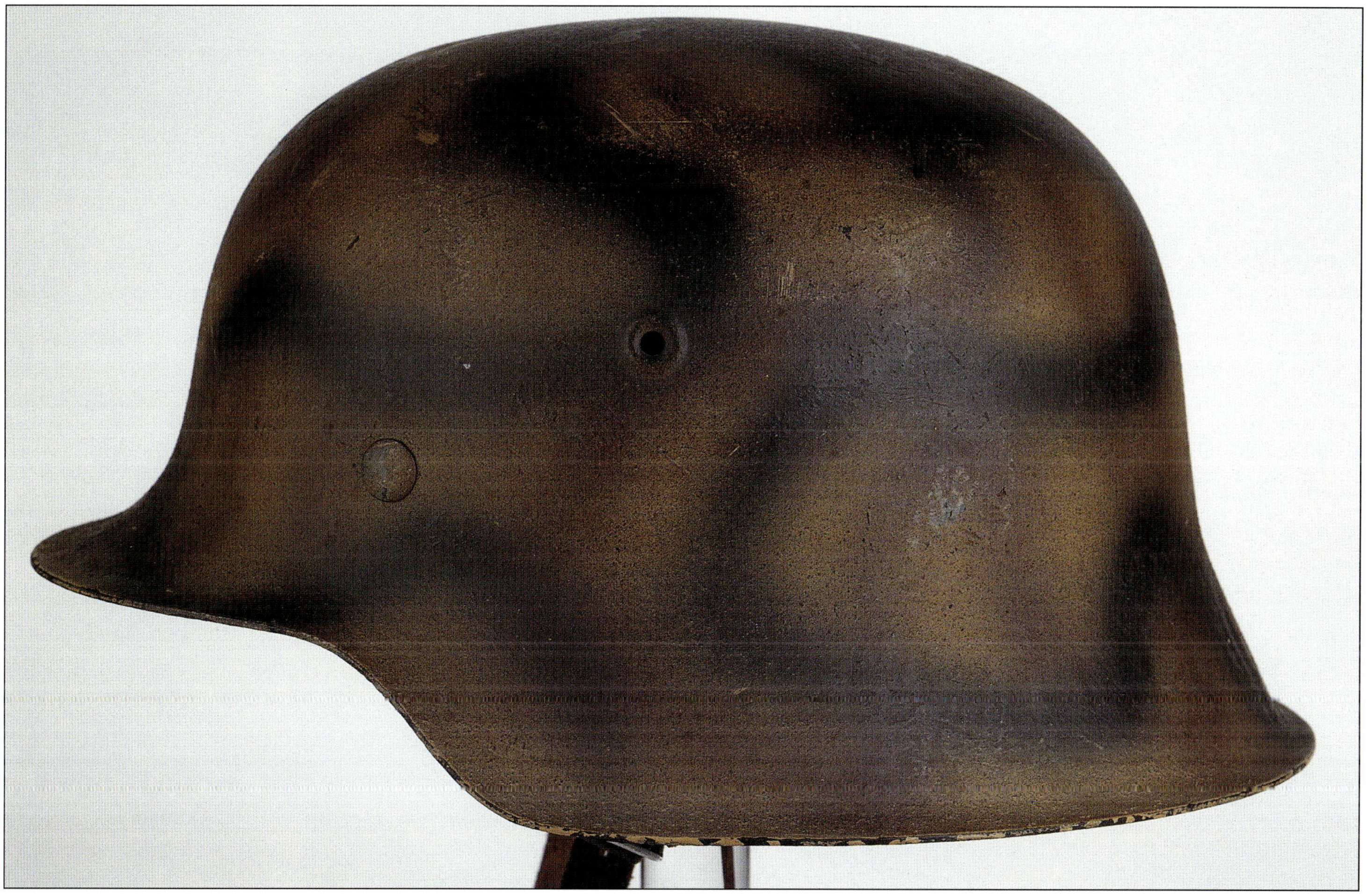

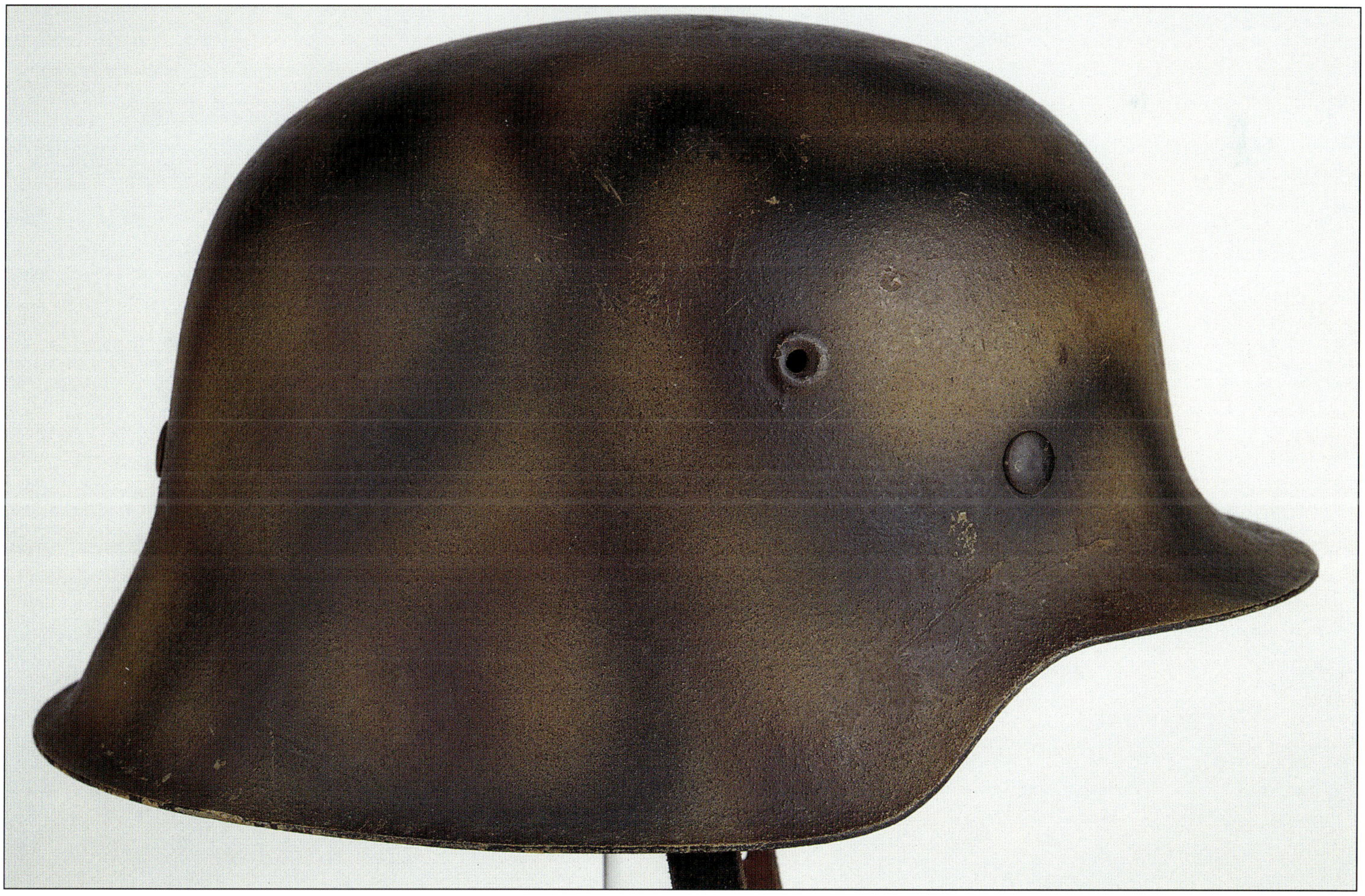

M42 helmet shell with M31 liner and chinstrap. This helmet was originally painted a smooth field grey and has been overpainted with a dark sand colour, then sprayed with patches of green. It carries the late pattern SS runic shield on the right side and has thick wire attached for foliage.

M42 helmet shell with M31 liner and chinstrap. This helmet was originally painted a smooth field grey and has been overpainted with a sand base colour, then lightly sprayed with patches of green and brown. It carries no insignia but has the name 'Str Metgeart' written on the liner, identifying it as an SS helmet.

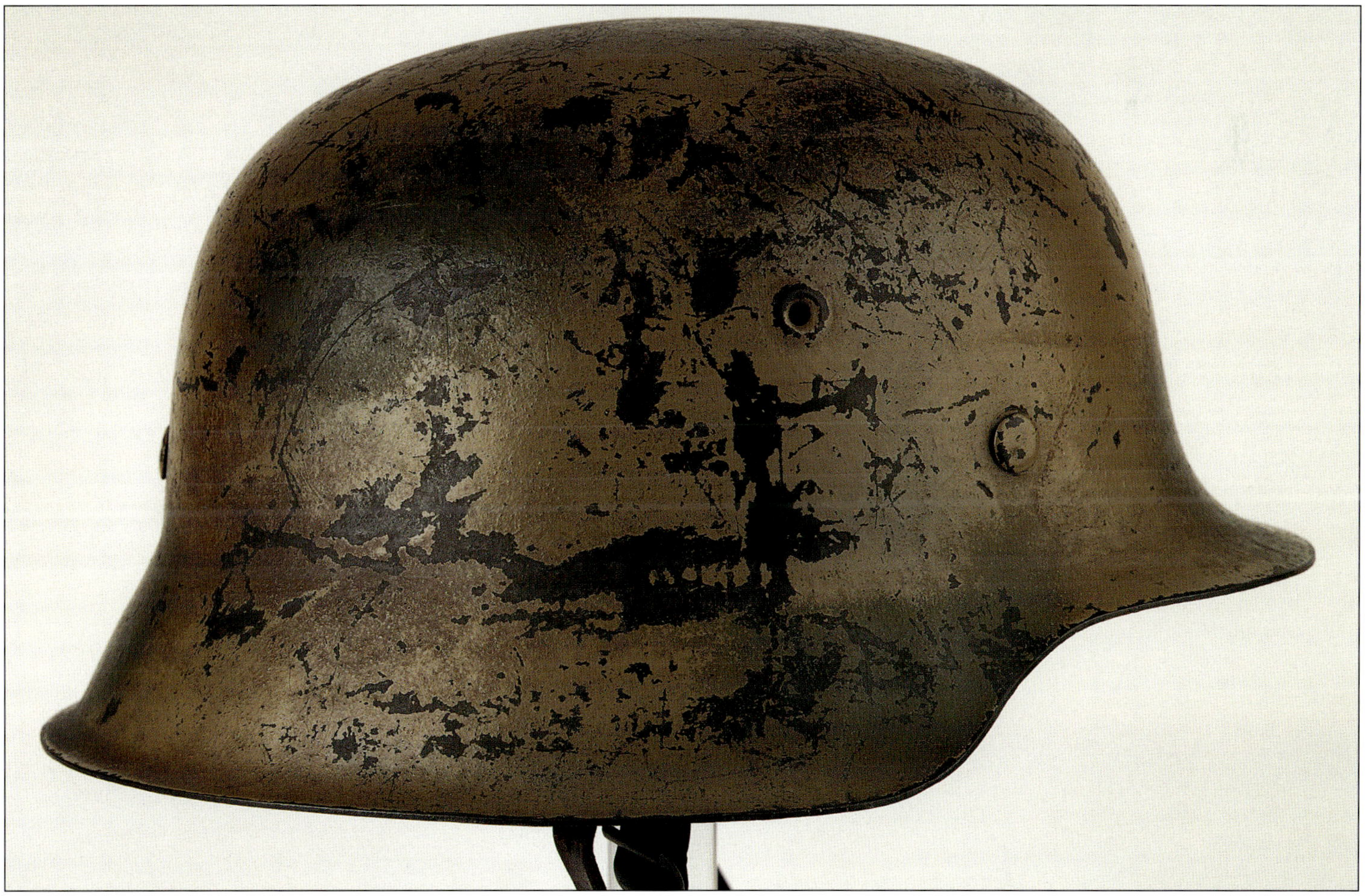

M42 helmet shell with M31 liner and chinstrap. This helmet was originally painted a matte field grey and has been over sprayed with white, brown and green. It carries the Kriegsmarine gold eagle transfer on the left side.

■ M42 helmet shell with M31 liner and chinstrap. This helmet was originally painted a smooth field grey and has been overpainted with a sand/orange colour, then brushed with green patches. It carries no insignia.

Interior detail shows the museum inventory ink stamp on the liner. This helmet originated from Czechoslovakia.

M42 helmet shell with M31 liner and chinstrap. This helmet was originally painted a matte field grey and has been covered with a zimmeritt textured paste, then painted a sand base colour with patches of green. No insignia is visible.

M42 helmet shell with M31 liner and chinstrap. This helmet is painted a matte field grey and carries no insignia. Chicken wire has been attached for foliage.

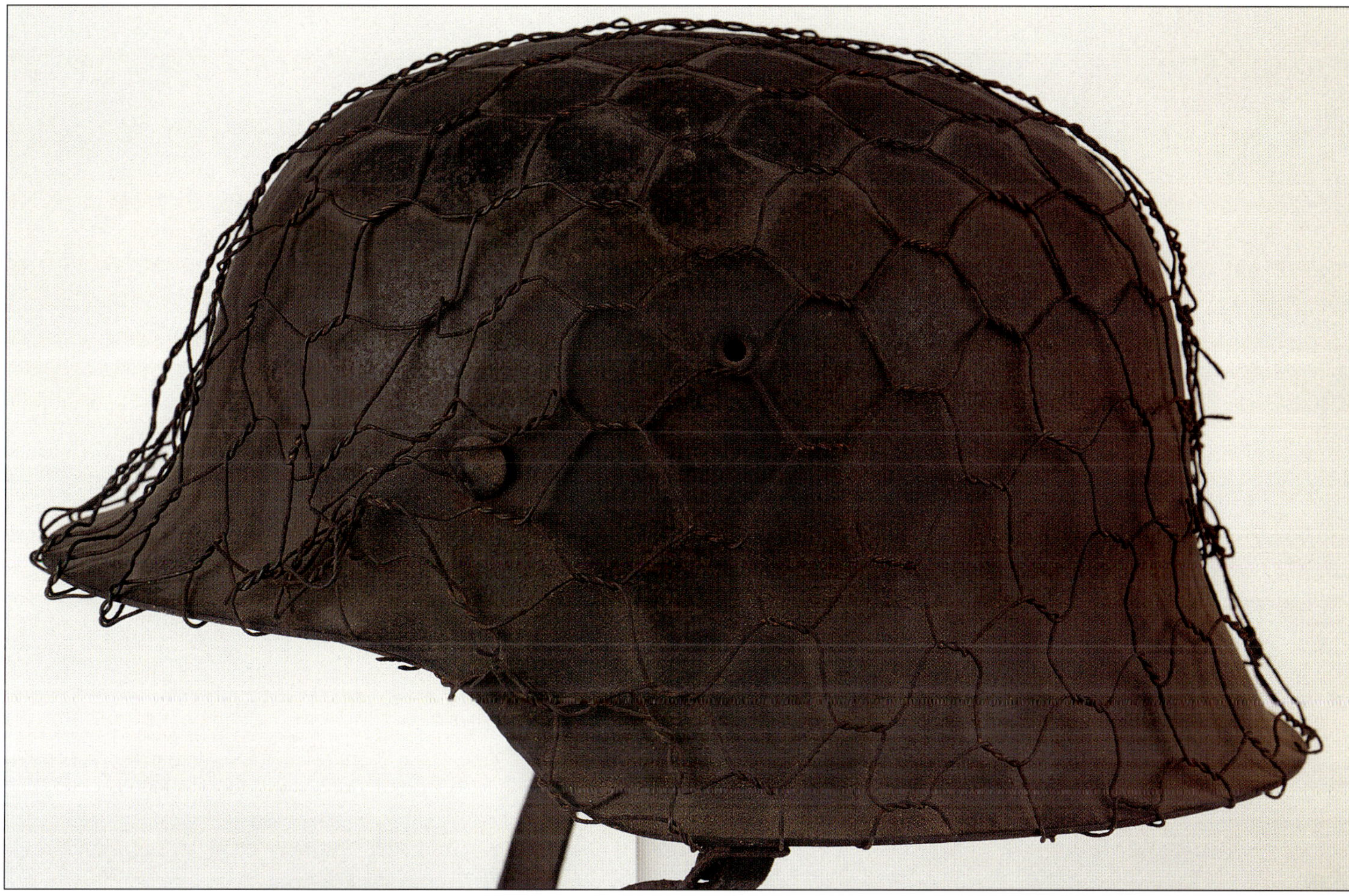

M42 helmet shell with M31 liner and chinstrap. This helmet is painted a matte field grey. It was originally an SS helmet with the runic shield on the right side, which has been removed. A Wehrmacht eagle transfer was placed on the left side. A heavy gauge wire and chicken wire covering is attached.

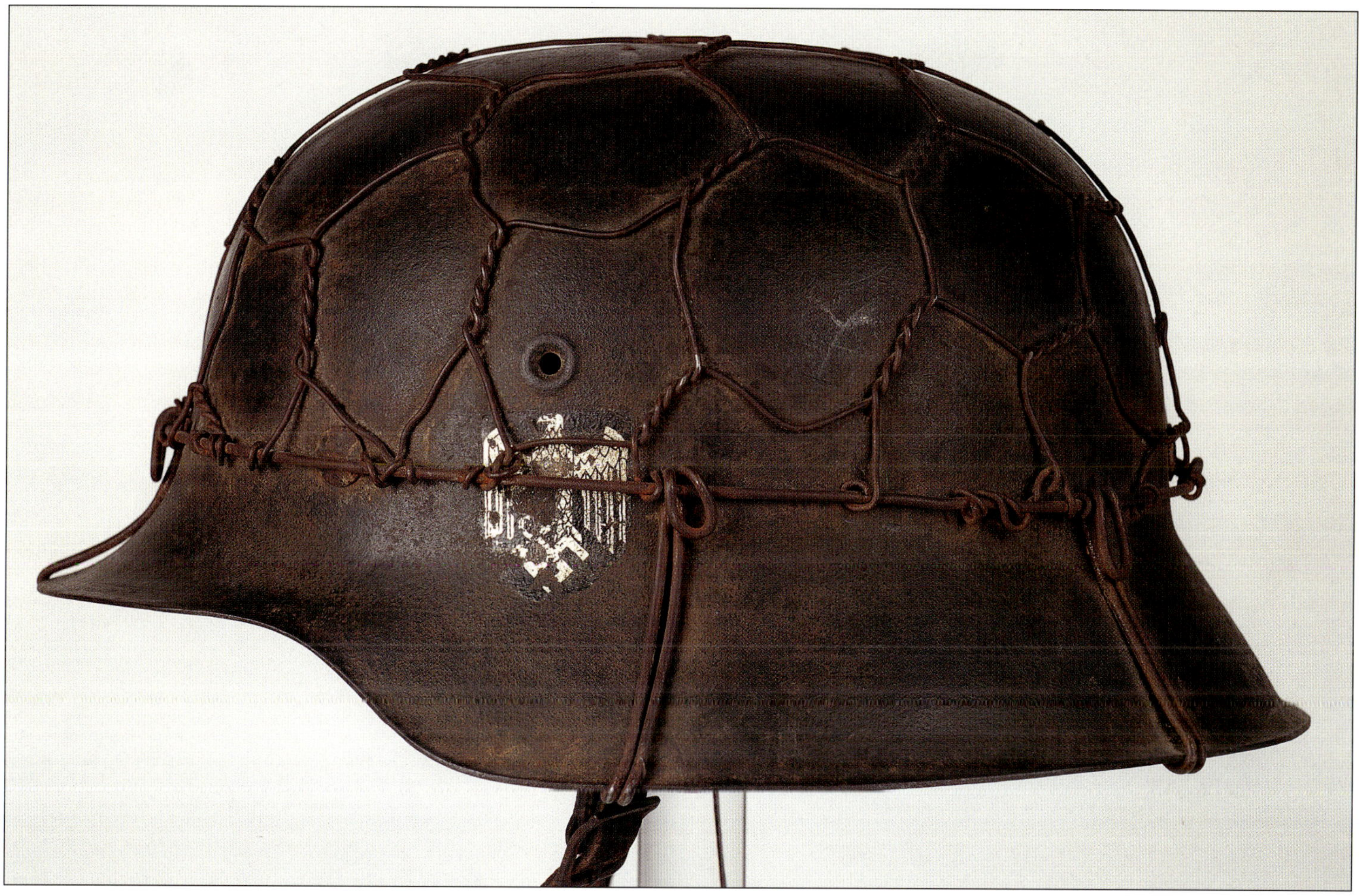

M42 helmet shell with M31 liner and chinstrap. This helmet is painted a matte field grey. It carries the late pattern SS runic shields on both sides and has a wire arrangement attached for foliage.

MODEL 1942

M42 helmet shell with M31 liner and chinstrap. This helmet was originally painted a matte field grey and has been overpainted with a base sand colour, then sprayed with red/brown and green patches. A late pattern Luftwaffe eagle transfer has been applied over the camouflage on the left side. A string netting cover with metal hooks has been attached for foliage. This pattern of netting cover is thought by some to be of post-war manufacture which may be the case with later versions. It is clear from the next photograph that the camouflage has been sprayed onto the helmet with the netting in place. Therefore, this particular netting cover was certainly produced before the end of the war.

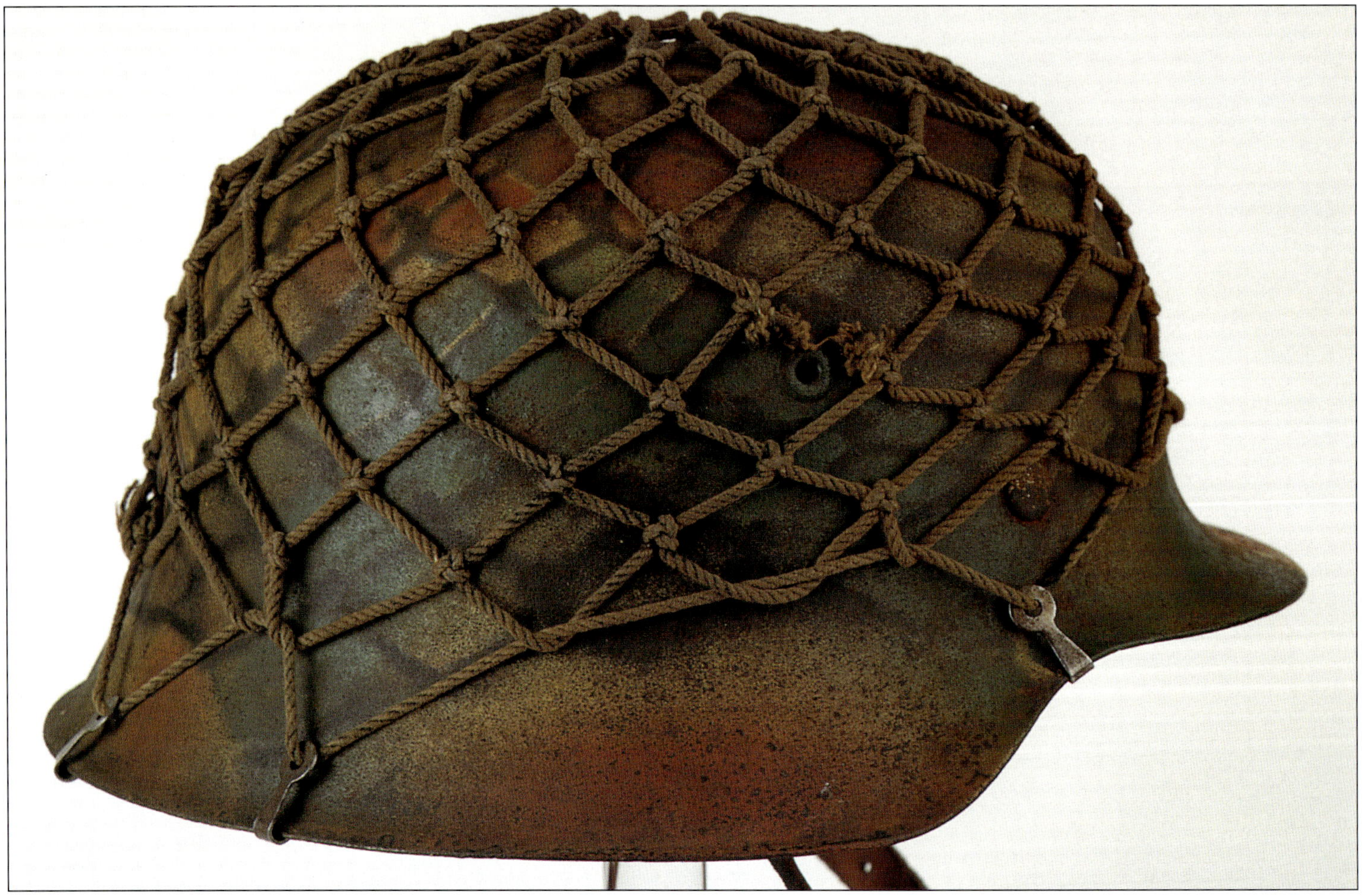

M42 helmet shell with M31 liner and chinstrap. This helmet was originally painted a matte field grey and has been covered with a zimmeritt textured paste, then painted with patches of light sand, brown and green. It has the full regulation string netting attached.

Frontal view with the face veil down. Notice the wire hooks which hold the veil when not required.

■ Rear view showing the attachment of the netting to the shell by a drawstring with a metal ring.

■ M42 helmet shell with M31 liner and chinstrap. This helmet is painted a matte field grey and has a field-made strap for foliage, taken from a bread bag, riveted and attached with hooks. It carries no insignia.

MODEL 1942/45

■ M42/45 helmet shell with M31 liner and chinstrap. This helmet is painted a satin field grey and carries the Wehrmacht eagle transfer on the left side. It appears to be of late-war manufacture as it is crudely formed and lacks the normal M40/42 ventilation holes. The liner conforms to the normal M31 configuration but the pinkish brown leather is of poorer quality.

NOTES

NOTES

The Companion Volume

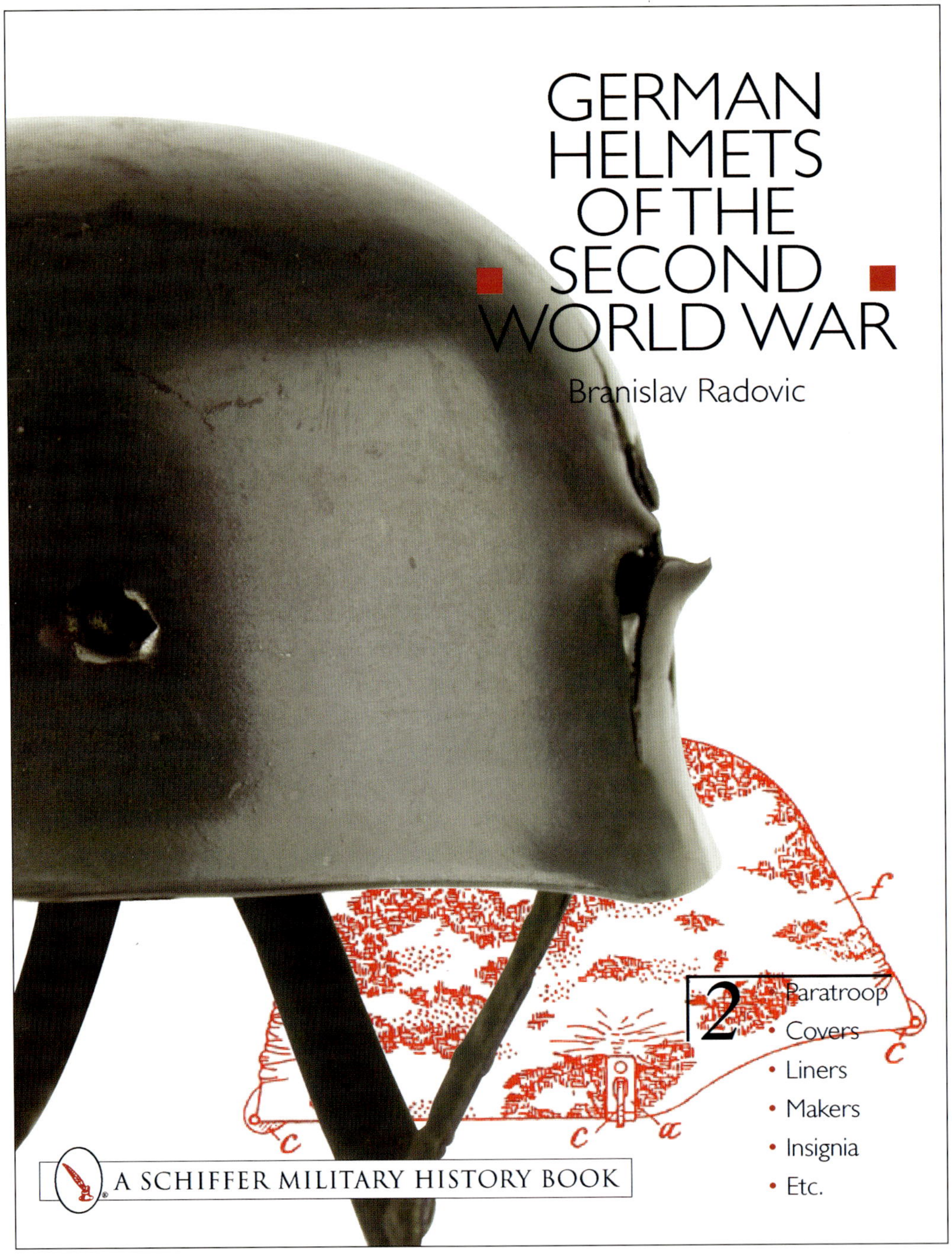

German Helmets of the Second World War

Volume Two

• Paratroop • Covers • Liners • Makers • Insignia • Etc. •

Branislav Radovic

Size: 9" x 12", over 480 color, and black and white photos
304 pages, hard cover
ISBN: 0-7643-1448-3 $79.95

Bonk